Lectures in Applied Mathematics

Volume 9
Lectures in Applied Mathematics

RELATIVITY THEORY AND ASTROPHYSICS

2. GALACTIC STRUCTURE

Jürgen Ehlers, EDITOR
The University of Texas

1967
American Mathematical Society, Providence, Rhode Island

The Summer Seminar was conducted, and the proceedings prepared in part, by the American Mathematical Society under the following contracts and grants:

Air Force Office of Scientific Research under Grant AF-834-65
Atomic Energy Commission under Contract AT(30)1364-2
National Aeronautics and Space Administration under Grant NsG 358
National Science Foundation under NSF Grant GE-7790
Office of Naval Research under Contract Nonr(G)00052-65.

Library of Congress Catalog Card Number 62-21481

Foreword

This volume is the second of a series of three which contain the Proceedings of the Fourth Summer Seminar on Applied Mathematics, arranged by the American Mathematical Society and held at Cornell University from July 26 to August 20, 1965.

The purpose of the Seminar was primarily to acquaint graduate students and recent recipients of the Ph.D. degree with the state of knowledge and current problems in Relativity and Astrophysics and thus to stimulate research in these subjects.

Because the participants could not be expected to be familiar with basic information in both the fields of relativity and astrophysics, five series of basic lectures were given. These consisted of ten lectures on the Theory of Relativity by A. Schild, four lectures on Theoretical Cosmology by E. Schücking, seven lectures on Galactic Structure and Galactic Dynamics by L. Woltjer, seven lectures on Stellar Structure by E. E. Salpeter, and six lectures on Stability Problems by S. Chandrasekhar. In addition, one or more lectures were given on Experimental Tests of General Relativity, Relativistic Hydrodynamics, Gravitational Collapse, Gravitational Radiation, Observational Cosmology, Cooperative Phenomena in Stellar Dynamics, Spiral Structure of Galaxies, and Cosmic Rays; a survey of the lectures is given in the Introduction contained in volume 8, the first of this series.

The program of the Seminar was organized by a committee consisting of the following members:

S. Chandrasekhar
C. C. Lin
C. W. Misner
A. Schild
A. H. Taub (Chairman).

Thanks are due to the five government agencies which supplied financial support, as acknowledged on the copyright page; to the staff of the American Mathematical Society for administrating the Seminar, particularly to Dr. Gordon L. Walker, Executive Director of the Society, who contributed greatly to the planning and functioning of the Seminar; to the members of the organizing committee listed above, especially its chairman; to Dr. William Smith, Director of Summer Session and Extramural Courses of Cornell University, and to Professor Martin Harwit, Department of Astronomy, for handling the housing and for the provision of other facilities for the Seminar and staff. The devotion of Mrs. M. L. Leigh to the discharging of her duties in the day to day operation of the Seminar office deserves special mention.

JÜRGEN EHLERS, EDITOR

THE UNIVERSITY OF TEXAS
AUSTIN, TEXAS

Contents

L. Woltjer

Structure and Dynamics
of Galaxies[1]

I. **Introduction.** A galaxy is a large, more or less self contained, system consisting of stars and gas. Masses range from 10^8 to 10^{12} solar masses, with $10^{11} \odot$ a typical value. Sizes range from probably much less than 10^3 light years for quasi-stellar galaxies to over 10^5 l.y., with several times 10^4 l.y. the most typical value. In most galaxies observed at present the mass of gas appears to be less than 20 per cent; in many cases a few per cent is more representative. But since short lived, bright, massive stars are formed in the gas, it may dominate the visual aspect of a galaxy. Most galaxies in our neighborhood seem rather old, about 10 billion years. Probably all galaxies started their existence as pure gas that gradually condensed into stars.

Galaxies mostly come in three distinct shapes: oblate spheroidal systems (elliptical or E-galaxies); flat, disk-like systems (S-galaxies); and nonaxisymmetric systems somewhat resembling prolate spheroids (SB-galaxies). There is some evidence that gas is rare in the E-galaxies. In the S and SB systems a characteristic spiral pattern, outlined mainly by gas and bright stars, is frequently in evidence.

[1] Supported in part by the National Science Foundation, GP 4194.

Since stars make up a large fraction of the total mass of most galaxies and since in some ways the dynamics of a stellar system is simpler than that of a gaseous one, we shall begin by developing a theoretical framework for the description of stellar systems. After this we shall return to the observed features of galaxies and to the role of the gas. At this point, however, it should already be clear that no full description of the evolution of galaxies can be given on the basis of a theory in which the dynamics of the gas is left unspecified.

Let us then consider a stellar system consisting of, say, 10^{11} stars. Each star moves in the gravitational field, produced by all the stars, in accordance with the equations of motion. Thus if the gravitational potential is $\Phi(x, y, z, t)$, the orbit of a star is given by

(1) $$d^2\mathbf{r}/dt^2 = - \nabla \Phi.$$

The potential is determined by Poisson's equation

(2) $$\nabla^2\Phi = 4\pi G\rho,$$

with ρ the total mass density and $G = 6.67 \times 10^{-8} \text{c}^3\text{g}^{-1}\text{s}^{-2}$ the gravitational constant.

Equations (1) and (2) in principle give a strict description of a self-gravitating stellar system. Only direct collisions between the stars are not taken into account, but for normal stellar systems these are so rare as to be completely negligible. Our description thus far, although exact, still is quite unwieldy. Strictly speaking ρ in Poisson's equation is a sum of 10^{11} δ-functions (if the stars are treated as point masses), while the equations of motion have to be written separately for the 10^{11} stars. A suitable statistical description thus is needed.

To begin with let us suppose that we write $\rho = \langle\rho\rangle + \delta\rho$, with $\langle\rho\rangle$ the spatially averaged mass density and $\delta\rho$ the fluctuating difference between ρ and $\langle\rho\rangle$. Similarly we split Φ into $\langle\Phi\rangle + \delta\Phi$, where $\langle\Phi\rangle$ is related to $\langle\rho\rangle$ by Poisson's equation. It now is our contention that in most stellar systems the dynamical effects of the fluctuating gravitational field are quite small compared to those of the mean field associated with the smoothed distribution of matter, at least during times of the order of the age of these systems.

Ia. *The time of relaxation.* Let us consider for simplicity a star (mass m, velocity V) that moves in a region where $\langle \Phi \rangle$ is constant. In the absence of the $\delta\Phi$'s associated with individual stars the orbit of the star then would be a straight line. We first estimate the deflection caused by an encounter with a single star of mass M, whose velocity we suppose to be equal to zero. If the impact parameter (the smallest distance between the unperturbed orbit and this star) is D, then roughly during a time $2D/V$ the moving star experiences a force GmM/D^2 transverse to the unperturbed orbit. Thus in the encounter the star acquires a velocity $\Delta v_\perp$ transverse to its original orbit given about by the product of the duration of the encounter and the acceleration,

$$(3) \qquad\qquad \Delta v_\perp \cong 2GM/DV.$$

It is easily verified that under the typical conditions listed in Table I in virtually all stellar encounters $\Delta v_\perp \ll V$. Thus we have to see how the effects of many small deflections have to be added. We *assume* that subsequent encounters are completely uncorrelated. Then since $\Delta \mathbf{v}_\perp$ is a vector whose direction (in the plane transverse to the unperturbed orbit) is random—because the positions of the stars that are encountered are random—it is clear that $\langle \Delta \mathbf{v}_\perp \rangle$ vanishes; but $\langle (\Delta v_\perp)^2 \rangle$ does not vanish. In fact, like in all stochastic processes we have after N encounters in each of which the magnitude of the deflection is $\Delta v_\perp$ that $\langle v_\perp^2 \rangle = N(\Delta v_\perp)^2$. The number of encounters with impact parameter between D and $D + dD$ in a time T is $2\pi TVnDdD$ when n is the density of the perturbing stars, which for simplicity are supposed not to move. The expectation value of $v_\perp^2$ after some time then is found by multiplying $(\Delta v_\perp)^2$ for a given D with the number of encounters during the time T with that impact parameter and subsequently integrating over all impact parameters. Thus

$$(4) \qquad \langle v_\perp^2 \rangle = \int_{D_{\min}}^{D_{\max}} 2\pi TVn \frac{4G^2M^2}{D^2V^2} DdD = \frac{8\pi G^2M^2Tn}{V} \ln \frac{D_{\max}}{D_{\min}}.$$

The integral diverges logarithmically for very large and for very small impact parameters. The difficulty at small impact parameters is only a consequence of our rough treatment and need not concern us here (cf. Appendix I). At large impact parameters the divergence is genuine. It has sometimes been supposed that $D_{\max}$ is to be taken equal to the mean interstellar distance, but more convincing arguments may be given for taking it equal to the size of the whole stellar system. In practice the difference is slight since D_{max} enters only logarithmically.

Much more significant than the uncertainties in D_{max} are the basic deficiencies in the theory. It has been assumed that the encounters can all be described as uncorrelated two-body encounters, and it is not yet clear how good this assumption is. To proceed beyond the two-body theory, however, would bring us into quite difficult, only partially understood problems in statistical mechanics.

We thus shall content ourselves with making use of Equation (4) where for $\ln \Lambda = \ln (D_{max}/D_{min})$ we shall take a representative value of 20. Clearly Equation (4) is only valid as long as $\langle v_\perp^2 \rangle \ll V^2$. When $\langle v_\perp^2 \rangle$ would become of the same order as V^2 the star would have completely lost its original direction, and its orbit would have been significantly affected by the $\delta\Phi$'s associated with individual stars. We thus define a relaxation time T_D (associated with the deflection of the orbit) as the time after which according to Equation (4) we would have $\langle v_\perp^2 \rangle = V^2$. Thus

$$(5) \qquad T_D = V^3 / (8\pi G^2 M^2 n \ln \Lambda) = V^3 / (8\pi G^2 M \rho \ln \Lambda),$$

and for times much smaller than T_D the effects of the $\delta\Phi$'s are generally small. Inserting numerical values and expressing the physical variables in convenient units we have

$$(6) \qquad T_D = 1 \times 10^8 V^3 (\text{km/sec}) / [M^2(\odot) n (\text{pc}^{-3})] \text{ years},$$

where $M(\odot)$ is the mass of the stars expressed in solar masses ($\odot = 2 \times 10^{33} g$) and $n(\text{pc}^{-3})$ is the number of stars per cubic parsec.[2] In this expression a correction for the motions of the field stars has been included, and the root mean square velocity of the stars was taken equal to V. Supposing also that $M = 1 \odot$ we obtain the results of Table I.

From the results in Table I it would appear that the deflections produced by the $\delta\Phi$'s are negligible in most galactic situations. In the small star clusters, however, relaxation effects will be quite important. The globular clusters are objects of more than 10 pc diameter and of the order of 10^5, solar mass like, stars. More than

---

[2] The parsec (pc or ps) is the standard unit of length in galactic studies. Originally being defined as the distance from which the mean radius of the orbit of the earth around the sun subtends an angle of one second of arc, it is equal to 3.086×10^{18} cm or to 3.1 light years. Note that to within 2 per cent 1 km/sec equals 1 pc/10^6 years.

TABLE I. The time of relaxation in some typical situations

Region	$n(\mathrm{pc}^{-3})$	$V(\mathrm{km/sec})$	$T_D(\mathrm{years})$
Solar neighborhood	0.1	10	10^{12}
Galactic center	10,000	200	10^{11}
Elliptical galaxy	10	200	10^{14}
Globular cluster			
inner region	1,000	10	10^{8}
outer region	1	10	10^{11}
Galactic cluster	10	0.5	10^{6}

100 such objects have been found in our galaxy. The galactic clusters are of the order of a few parsec in linear dimension and contain a few thousand solar masses of stellar matter, frequently including stars of 10—50 solar masses.

In addition to the time T_D we can define relaxation times associated with the exchange of energy in stellar encounters. Since these times turn out to be similar to T_D we defer further discussion to Appendix I.

One basic uncertainty still remains in our estimates in Table I. The relaxation time depends on the mass of the perturbing stars even for a given density of stellar matter. Thus if a galaxy—the mean density of which can frequently be estimated rather well from dynamical data—consisted not of single stars, but of small clusters of stars, the relaxation times could be significantly reduced. And it would be very difficult to detect small clusters in most galaxies. Also even transient clusters could have these effects.

If we neglect the possibilities of clustering, however, it appears that in the dynamics of galaxies the relaxation effects are small. Thus we will be justified in inserting the smoothed distribution of matter in Poisson's equation and to make use of the resulting Φ in the equations of motion.

II. **The Boltzmann equation.** We thus begin by assuming that the effects of encounters between stars are negligible and that consequently each star moves in a reasonably smooth gravitational field according to the equation of motion

(7) $$d\mathbf{p}/dt = m\,d\mathbf{v}/dt = -m\,\nabla\,\Phi.$$

Consider a six-dimensional phase space in which the coordinates are the three spatial coordinates (q_1, q_2, q_3) and the three momenta (p_1, p_2, p_3). Each star can, at a given instant, be characterized by a representative point in phase space. The representative point moves along a trajectory in phase space in accordance with the equation of motion in which $\Phi(q_1, q_2, q_3, t)$ should be considered as a known function of its arguments. Suppose that a large number of stars are considered and that in phase space a density of representative points $\Psi(\mathbf{q}, \mathbf{p}, t)$ can be meaningfully defined. Here and in the following $\mathbf{q}$ stands for (q_1, q_2, q_3) as argument, and $d\mathbf{q}$ will sometimes be used as an abbreviation for dq_1, dq_2, dq_3.

We derive a continuity equation for Ψ. Consider a small box enclosed by the hypersurfaces $q_1 = \text{const}$, $q_1 + dq_1 = \text{const}$, $\cdots, p_3 + dp_3 = \text{const}$. The volume of this box is $d\tau = dq_1 dq_2 dq_3 dp_1 dp_2 dp_3$. Consider the five-dimensional hypersurface dS_5 of the box, characterized by $q_1 = \text{constant}$. Through this hypersurface, which is orthogonal to the q_1-axis, $(\Psi \dot{q}_1)_{q_1} dS_5$ representative points enter the box per unit time, while at the opposite hypersurface $(\Psi \dot{q}_1)_{q_1 + dq_1} dS_5$ representative points leave. Here $\dot{q}_1 = dq_1/dt$ is the velocity of the representative points in the q_1 direction. Since dq_1 is infinitesimal, $(\Psi \dot{q}_1)_{q_1 + dq_1} = (\Psi \dot{q}_1)_{q_1} + (\partial \Psi \dot{q}_1/\partial q_1)_{q_1} dq_1$, if Ψ is a well-behaved function. Thus the net increase per unit time in the number of particles in the box due to the flow through the two hypersurfaces orthogonal to q_1 is $- (\partial \Psi \dot{q}_1/\partial q_1) \, dq_1 dS_5 = - (\partial \Psi \dot{q}_1/\partial q_1) \, d\tau$. To find the total increase in the number of points in the box, $(\partial \Psi/\partial t) \, dt$, we sum over all six pairs of hypersurfaces and obtain (summation from 1 to 3 over repeated indices)

$$(8) \qquad \partial \Psi/\partial t = - [\partial (\Psi \dot{q}_i)/\partial q_i + \partial (\Psi \dot{p}_i)/\partial p_i].$$

This is of course the six-dimensional analogue of the ordinary hydrodynamical equation of continuity $\partial \rho/\partial t = - \nabla \cdot (\rho \mathbf{u})$. Clearly $m \partial \dot{q}_i/\partial q_i = \partial p_i/\partial q_i = 0$, as p_i and q_i are independent coordinates. From the equation of motion it follows that $\dot{p}_i$ depends on q_i (through Φ), but not on p_i, and thus $\partial \dot{p}_i/\partial p_i = 0$. Thus

$$(9) \qquad D\Psi/Dt = \partial \Psi/\partial t + \dot{q}_i \partial \Psi/\partial q_i + \dot{p}_i \partial \Psi/\partial p_i = 0.$$

Thus the representative points move through phase space like an incompressible fluid, and Ψ is constant along a trajectory (Liouville's theorem).

A more general derivation is obtained by taking q_1, q_2, q_3 and

p_1, p_2, p_3 to represent a general set of Hamiltonian coordinates and their conjugate momenta. Then if the Hamiltonian is H we have the equations of motion $\dot{p}_i = -\partial H/\partial q_i$ and $\dot{q}_i = \partial H/\partial p_i$, and therefore $\partial \dot{q}_i/\partial q_i + \partial \dot{p}_i/\partial p_i = 0$. Then we have

(10) $\qquad \partial \Psi/\partial t + (\partial H/\partial p_i)(\partial \Psi/\partial q_i) - (\partial H/\partial q_i)(\partial \Psi/\partial p_i) = 0.$

This equation is independent of the coordinates chosen. We make use of it to obtain the equivalent of Equation (9) in cylindrical coordinates. In that case $q_1 = \varpi$, $q_2 = \phi$, $q_3 = z$, and the conjugate momenta are $p_\varpi = m\dot{\varpi}$, $p_\phi = m\varpi^2\dot{\phi}$, $p_z = m\dot{z}$, while

$$H = \frac{1}{2m}\left(p_\varpi^2 + \frac{1}{\varpi^2}p_\phi^2 + p_z^2\right) + m\Phi.$$

Inserting this in Equation (10) we have

(11)
$$\frac{\partial \Psi}{\partial t} + \frac{p_\varpi}{m}\frac{\partial \Psi}{\partial \varpi} + \frac{p_\phi}{m\varpi^2}\frac{\partial \Psi}{\partial \phi} + \frac{p_z}{m}\frac{\partial \Psi}{\partial z} - \left(m\frac{\partial \Phi}{\partial \varpi} - \frac{p_\phi^2}{m\varpi^3}\right)\frac{\partial \Psi}{\partial p_\varpi}$$
$$- m\frac{\partial \Phi}{\partial \phi}\frac{\partial \Psi}{\partial p_\phi} - m\frac{\partial \Phi}{\partial z}\frac{\partial \Psi}{\partial p_z} = 0.$$

In Cartesian coordinates the distribution function $f(\mathbf{r}, \mathbf{v}, t)$ which represents the number of stars in an element of volume with coordinates between $\mathbf{r}$ and $\mathbf{r} + d\mathbf{r}$ and velocities between $\mathbf{v}$ and $\mathbf{v} + d\mathbf{v}$ at the time t can be directly identified with the density of representative points in phase space, Ψ, if the momenta are replaced by velocities. Thus (with the x_i the ordinary spatial coordinates)

(12) $\qquad \partial f/\partial t + v_i \partial f/\partial x_i - (\partial \Phi/\partial x_i)(\partial f/\partial v_i) = 0.$

This is the collision-free Boltzmann equation, sometimes referred to as the "The equation of continuity," the "Vlasov equation," or the "Liouville equation."

For a self-gravitating system this equation is to be supplemented by Poisson's equation, which is now written as

(13) $\qquad\qquad \nabla^2 \Phi = 4\pi Gm \int f\, d\mathbf{v},$

where the integration is over all velocity space and where all stars are supposed to have mass m. If stars with different masses occur, separate distribution functions f_m must be introduced. Each f_m must satisfy the same Boltzmann Equation (12), while on the right-hand side of Equation (13) mf is replaced by $\sum mf_m$. We shall not consider such composite cases.

The determination of f thus generally involves the simultaneous solution of Equations (12) and (13). The problem is clearly a non-linear one. A much simpler problem occurs when we consider the dynamics of a rather scarce stellar population whose contribution to the total mass density is negligible. Then Φ in Equation (12) is a given function, and only the linear Equation (12) has to be solved.

IIa. *Stationary solutions and integrals of motion.* Most stellar systems that we can analyze in detail appear to have ages of the order of 10^{10} years. Since individual stars move through these systems in times of the order of 10^8 years, it seems reasonable to suppose that some kind of a steady state has been reached. We thus will consider steady stellar systems described by the time independent Boltzmann equation

(14) $v_i \partial f/\partial x_i - (\partial \Phi/\partial x_i)(\partial f/\partial v_i) = 0.$

Let us for the moment consider Φ as given. Then Equation (12) is a homogeneous linear equation for $f(x, y, z, v_x, v_y, v_z, t)$, and thus we can write (A. G. Webster, *Differential equations of physics*, Stechert, New York, 1933, pp. 59, 60)

(15) $f = f(I_1, I_2, \cdots, I_6),$

where $I_1, \cdots, I_6$ are the six integrals of the Lagrangian subsidiary equations

(16) $dt = \dfrac{dx}{v_x} = \dfrac{dy}{v_y} = \dfrac{dz}{v_z} = -\dfrac{dv_x}{\partial \Phi/\partial x} = -\dfrac{dv_y}{\partial \Phi/\partial y} = -\dfrac{dv_z}{\partial \Phi/\partial z},$

that is, of the equations of motion for single particles. Of course, the determination of the integrals may be as difficult as the solution of the original equation.

If stationary solutions are sought the time can be eliminated between the six integrals, and f depends on five independent integrals. The physical situation is clear. Consider the case where only one integral I_1 would be present. Then the trajectory of a particle would be situated in phase space on the hypersurface $I_1 = $ const., and nothing would prevent the trajectory to fill this hypersurface ergodically. According to Liouville's theorem, in a steady state the distribution function is constant along a trajectory in phase space, and thus if the trajectory fills the hypersurface, f should be constant on it, but not necessarily on different hypersurfaces, since no connecting trajectories can occur. Thus we would have $f = f(I_1)$. If two integrals are present then f would be constant on each hypersurface given by $I_1 = $ const., $I_2 = $ const., and thus $f = f(I_1, I_2)$. The intersection of all hypersurfaces $I_i = $ const. just gives the trajectory of a particle through phase space. If all integrals entered in the same way the present method of solution would not be very useful, as a complete description of the system would be required. But this is not the case. Not all integrals isolate points on the trajectory from other points in phase space. Conservative (that is, not explicitly time dependent) integrals which have this property are said to be isolating. If, for example, the surface I $= $ const. is infinitely multiple-valued, the integral may not isolate.

An example of a two-dimensional (x, y) harmonic oscillator will illustrate the situation. Suppose the potential is $\Phi_0 + \alpha^2 x^2/2 + \beta^2 y^2/2$. We have

(17)
$$\ddot{x} + \alpha^2 x = 0; \qquad x = A \sin \alpha (t - t_1),$$
$$\ddot{y} + \beta^2 y = 0; \qquad y = B \sin \beta (t - t_2).$$

Thus the solution is periodic if α is a rational fraction of β. If not, the trajectory fills the rectangle $x^2 < A^2$, $y^2 < B^2$ ergodically.

Let us now examine the problem from the integral point of view. Two integrals are written down immediately. After multiplying Equations (17) with $\dot{x}$ and $\dot{y}$ we have

(18)
$$\dot{x}^2 + \alpha^2 x^2 = I_1, \qquad \dot{y}^2 + \beta^2 y^2 = I_2.$$

$I_1 + I_2$ is the energy integral. Eliminating t between x and y in Equation (17) we have for the third integral

$$I_3 = \frac{1}{\alpha} \sin^{-1} \left(\frac{x}{A} \right) - \frac{1}{\beta} \sin^{-1} \left(\frac{y}{B} \right)$$
$$= \frac{1}{\alpha} \sin^{-1} \left(\frac{\alpha x}{I_1^{1/2}} \right) - \frac{1}{\beta} \sin^{-1} \left(\frac{\beta y}{I_2^{1/2}} \right)$$

If α/β is rational the integral is isolating, if not the integral is not isolating, and the orbit fills the possible rectangle ergodically. For consider a definite value of I_3 and of y. We have

(20)
$$x = \frac{I_1^{1/2}}{\alpha} \sin \left[\alpha I_3 + \frac{\alpha}{\beta} \mathrm{Sin}^{-1} \left(\frac{\beta y}{I_2^{1/2}} \right) + \frac{\alpha}{\beta} n\pi \right],$$

where now the Sin^{-1} refers to the principal value and where n is an integer. If α/β is rational, only a finite number of values of x are possible. But if it is irrational all n correspond to different values of x, and any x can be approached arbitrarily closely. I_1 and I_2 are isolating integrals. They define the rectangle within which the orbit should be contained as $x^2 < I_1/\alpha^2$ and $y^2 < I_2/\beta^2$. I_1, I_2, I_3 are all the conservative integrals as phase space is four-dimensional in the present case.

It is thus clear that the number of isolating integrals depends on the structure of the potential, and we must investigate the integrals which should be considered in galactic dynamics. If the potential is time independent, the energy E is an isolating integral. We have

(21) $$E = \tfrac{1}{2} v^2 + \Phi.$$

But to obtain additional integrals more specific assumptions must be made about the potential, and we discuss some typical cases.

IIb. *Spherical systems.* If we consider a stellar system with a spherically symmetrical density distribution the potential will be $\Phi(r)$ if we analyze the problem in spherical polar coordinates (r, ϕ, θ). Clearly in this potential the angular momentum vector of every stellar orbit is a constant of the motion, and we have immediately four isolating integrals E and $\mathbf{J}$.

Consider for example a star moving in an inverse square field. Then the orbit is of course an ellipse. From E and $\mathbf{J}$ we can find the orbital plane and the size and shape of the orbit. A fifth and last integral is needed to specify the orientation of the orbit in the orbital plane. In an inverse square field the orientation of the elliptical orbit remains fixed and thus the fifth integral is also isolating. But if a more general potential field is taken the orbit will no longer be an ellipse, but rather will be of the rosette type, and in this case the fifth integral is not isolating and the orbit fills the allowed region of the orbital plane ergodically. Thus in a spherical system we can write

(22) $$f = f(E, \mathbf{J}),$$

with f an arbitrary (positive definite) function of the arguments, as the general solution of the Boltzmann equation, if pathological cases like $\Phi \propto r^{-1}$ are excluded. In a real system such special potentials cannot be expected to be realized exactly.

We now introduce the solution (22) into Poisson's equation which enables us to obtain f explicitly as a function of the spatial and velocity variables. We have

$$(23) \qquad \nabla^2\Phi = 4\pi Gm \int f(\tfrac{1}{2}v^2 + \Phi, \mathbf{r}\times\mathbf{v})\,d\mathbf{v},$$

which after the integration over the velocity variables has been performed (for a definite function f) is an ordinary differential equation for Φ. It is to be solved with the boundary conditions: $d\Phi/dr = 0$ at $r = 0$ and $\Phi \propto 1/r$ if $r \to \infty$.

As an example we consider a very simple distribution function:

$$(24) \qquad f = g\exp(-2\alpha^2 E) = g\exp(-\alpha^2(v^2 + 2\Phi)).$$

Then Poisson's equation gives

$$(25) \qquad \begin{aligned} \nabla^2\Phi &= 16\pi^2 Gmg\exp(-2\alpha^2\Phi)\int_{-\infty}^{\infty}\exp(-\alpha^2 v^2)\,v^2 dv \\ &= \frac{4\pi^{5/2}Gmg}{\alpha^3}\exp(-2\alpha^2\Phi) \end{aligned}$$

which is the equation for a self-gravitating isothermal gas sphere. From this equation $\Phi(r)$ is easily obtained by numerical integration. An isothermal gas sphere has infinite radius and mass for finite central density. To obtain a finite system a cutoff must be applied to the velocity distribution, which eliminates stars with velocities in excess of the velocity of escape ($V_e^2 = -2\Phi$).

Next we consider a more complicated f like

$$(26) \qquad f = g\exp(-2\alpha^2 E - \beta^2 J^2).$$

In this distribution function stars with large angular momenta are suppressed compared to the previous case, and at large distances from the center most orbits must be almost radial. Writing in Equation (26) $J^2 = r^2 v^2 \sin^2\psi$ with ψ the angle between $\mathbf{r}$ and $\mathbf{v}$ we easily verify that the density becomes

$$(27) \qquad n = (\pi^{3/2}g/\alpha(\alpha^2 + \beta^2 r^2))\exp(-2\alpha^2\Phi),$$

which when inserted in Poisson's equation enables us to obtain $\Phi(r)$. With this distribution function we obtain for the ratio of the mean square radial velocity to the mean square total random velocity

(28)
$$\frac{\langle v_r^2 \rangle}{\langle v^2 \rangle} = \frac{\int f v_r^2 d\mathbf{v}}{\int f v^2 d\mathbf{v}} = \frac{1}{3 - 2\beta^2 r^2/(\alpha^2 + \beta^2 r^2)}.$$

Thus at $r = 0$ the velocity distribution is isotropic, while if $r \to \infty$ only the radial component survives. Such a distribution function— with suitable energy cutoff—might be appropriate for a globular cluster, if the stars in the outer part were originally formed in the inner region of the cluster and subsequently brought into more eccentric orbits by the effects of stellar encounters.

We note that the density is obtained by integrating f over all velocity space. Thus any part of f that is odd in the velocities does not contribute to the density. Such an odd part corresponds, however, to systematic mass motions. Thus, as pointed out by Lynden-Bell, a spherical cluster may well be rotating. There the difference with collisionally relaxed systems, where rotation necessarily leads to flattening, is conspicuous. Of course since stellar systems originate from gaseous configurations one still would expect that usually the systematic angular momentum will affect the shape of the cluster.

This discussion also shows the limitations of the equilibrium theory for collision-free systems: The functional dependence of f on its arguments is arbitrary. It can only be obtained on the basis of evolutionary considerations. But in such considerations one cannot avoid discussing the physics of the gaseous protogalaxy and the process of star formation.

Some of the arbitrariness of f might be removed by stability considerations. Clearly f's leading to instable systems must be rejected. Since some aspects of the stability of stellar systems are discussed in Lynden-Bell's paper of these Lectures, we shall not further consider this topic.

IIc. *Axisymmetric potentials*; *applications to the Galaxy*. In the spherical case we could find all isolating integrals without further knowledge about the structure of the potential. Also, Poisson's equation reduced to an ordinary differential equation can be solved comparatively easily. In even the simplest nonspherical cases it becomes impossible to find all isolating integrals with certainty, without making rather drastic assumptions, while Poisson's equation becomes a partial nonlinear differential equation.

If the potential is nonspherical but axisymmetric, the component of angular momentum along the symmetry axis is clearly conserved. Thus in addition to E we have an isolating integral J_z (in cylindrical coordinates ϖ, ϕ, z). Let us now consider a thin disk-like galaxy. In the neighborhood of the plane of symmetry we can approximately separate the potential as

(29) $$\Phi(\varpi, z) = \Phi_1(\varpi) + \Phi_2(z),$$

if not too large regions are considered. In this case the motions in the plane and transverse to it are uncoupled, and the energies associated with each of these motions are separately conserved. Thus in addition to the total energy E and the angular momentum J_z we have a third isolating integral

(30) $$I_3 = \tfrac{1}{2} v_z^2 + \Phi_2(z).$$

It can be shown that in this case no other isolating integral exists, corresponding to the absence of a fixed orbital plane. It would appear that the potential (29) can approximately represent conditions in the solar neighborhood, if very low velocity stars are considered. The high velocity stars roam through such large regions of the galaxy that the representation (29) is certainly invalid. The distribution function for the solar neighborhood stars thus would be

(31) $$f = f(E, J_z, I_3) = f\left[\tfrac{1}{2} v_\varpi^2 + \tfrac{1}{2} v_\phi^2 + \tfrac{1}{2} v_z^2 + \Phi, \varpi v_\phi, \tfrac{1}{2} v_z^2 + \Phi_2(z)\right].$$

If for a moment we would take f independent of I_3 then f would be the same function of v_ϖ as of v_z. Therefore all quantities relating to v_ϖ and v_z should be the same. In particular we would expect that the velocity dispersions $\langle v_\varpi^2 \rangle$ and $\langle v_z^2 \rangle$ should be identical. Observations show, however, that approximately $\langle v_\varpi^2 \rangle \cong 4 \langle v_z^2 \rangle$, which clearly establishes the importance of I_3.

About the same relation holds for the velocity dispersions of high velocity stars. This indicates that an isolating I_3 exists also for potentials of the galactic type more general than the potential (29). It is known that potentials that are separable in spheroidal coordinates admit an isolating I_3, but again there is no reason why in a real galaxy the potential should be precisely of this type. Numerical calculations of stellar orbits by Contopoulos, Ollongren, Torgård and others give rather convincing evidence that an I_3 is almost always present in a force field of galactic type. Unfortunately the integral cannot be written down a priori, and thus the construction of joint solutions of the Boltzmann and Poisson equations with an I_3 taken into account poses an awkward problem. It may well be that the numerical problems involved in the construction of such solutions are so formidable, that a numerical evolutionary calculation—which would be much more informative—might be more worthwhile. We shall not discuss the problems of the third integral here further since these matters are discussed in Contopoulos' paper of these Lectures.

A second application of I_3 in the solar neighborhood is of importance. The distribution of z-velocities of the stars near the sun is observed to be roughly Maxwellian in a first approximation. If this were strictly the case we would infer that

(32)
$$f(E, J_z, I_3) = f_1(E - I_3, J_z) \exp(-2\alpha^2 I_3)$$
$$= f_1(E - I_3, J_z) \exp(-\alpha^2 v_z^2) \exp(-2\alpha^2 \Phi_2(z)),$$

with f_1 independent of z and v_z. Hence

(33)
$$f(z)/f(0) = \exp\{-2\alpha^2[\Phi_2(z) - \Phi_2(0)]\}.$$

Thus in this case the velocity dispersion is independent of the altitude above the galactic plane, and the density law for a group of stars is that for an isothermal atmosphere. Of course if the distribution of stellar velocities in the z-direction is not Maxwellian we can always represent it as a sum of Maxwellian distributions and take the proper summation to find the density distributions. Then the components of highest velocity dispersion will fall off most slowly, and the velocity dispersion will increase with height. In the solar neighborhood a representation with two or three Maxwellian distributions is quite adequate. Restricting ourselves to one Maxwellian distribution, we have after taking the logarithm of Equation (33) and differentiating twice

(34)
$$-\frac{\partial^2 \Phi}{\partial z^2} = \frac{1}{2\alpha^2} \frac{\partial^2 \ln f(z)}{\partial z^2} = \langle v_z^2 \rangle \frac{\partial^2 \ln n(z)}{\partial z^2}.$$

Thus if for some clearly recognizable stellar population we can determine the velocity dispersion and the second z-derivative of the density n (twice differentiating observed functions of course always greatly magnifies errors in the observations) we can find $\partial^2 \Phi / \partial z^2$ near the sun. As we shall discuss later, measurements of galactic rotation enable us to find $(1/\varpi) \, \partial(\varpi \, \partial \Phi/\partial \varpi)/\partial \varpi$ with good accuracy. In fact, if we neglect pressure effects and magnetic effects, the galactic gas layer would be in centrifugal equilibrium, that is, each element of gas would rotate at such a speed that the centrifugal acceleration in its circular orbit θ^2/ϖ—with θ the circular velocity—would exactly balance the gravitational acceleration $-\partial\Phi/\partial\varpi$. Thus we would have

$$(35) \qquad \Theta^2/\varpi = \partial\Phi/\partial\varpi$$

and hence

$$(36) \qquad \frac{1}{\varpi}\frac{\partial}{\partial\varpi}\left(\varpi\frac{\partial\Phi}{\partial\varpi}\right) = \frac{2\Theta}{\varpi}\frac{\partial\Theta}{\partial\varpi}.$$

Obtaining $\Theta(\varpi)$ from observation and adding Equations (34) and (36) we obtain $\nabla^2\Phi$ and thus from Poisson's equation the total density of matter near the sun. In this dynamical determination of the density the term $\partial^2\Phi/\partial z^2$ is the largest, and it is responsible for the uncertainty of the final result.

It should be noted that in applying Equation (34) it is sufficient to consider only one kind of star, but that the application of Poisson's equation yields the density of all matter, stellar and gaseous. The recent determination of Hill and Oort leads to a mass density of $0.15\,\odot\,\mathrm{pc}^{-3}$ or about $1 \times 10^{-23}\,\mathrm{g/cm^3}$ with a stated uncertainty of about 10 percent. It is of interest to compare this value with the observed densities of stars and gas near the sun. It is seen from Table II that only about 60 percent of all the mass is accounted for. The balance might be made up of molecular hydrogen, which cannot be observed from under the earth's atmosphere, or by very faint low mass stars. In the former case the amount of gas would be more than tripled, in the latter the low mass stars should be quite numerous since they would contribute as much as all other stars together. It can be expected that observational evidence on this matter will become available in the near future.

TABLE II. Mass densities
of some galactic constituents near the sun

Constituent	density in $\odot/\mathrm{pc}^3$
Atomic + ionized hydrogen	0.025
Main sequence stars	0.049
Dark companions from binary stars } White dwarfs }	0.013
Total observed	0.087
Dynamical determination	0.150

We have by now arrived at a semiempirical level, but before proceeding farther let us return for a moment to our original objective of finding exact solutions of the Boltzmann and Poisson equations. We have already noted that in general I_3 cannot be explicitly written down. But if we restrict ourselves to distribution functions $f = f(E, J_z)$ we still have valid solutions of the Boltzmann equation, although not the most general solutions. Prendergast has recently begun the construction of complete galactic models— using in a first attempt $f \propto \exp(-\alpha^2 E - \beta J_z)$ with a suitable cutoff—and the first results of the numerical integration of the resulting Poisson equation appear very promising. Such models would be particularly suitable for the analysis of elliptical galaxies.

Thus for spherical galaxies the construction of equilibrium models poses no special problems. For elliptical or disk-like systems rather wide classes of models can be obtained. But for the barred SB systems no suitable models are in sight at the moment. Since these objects are not axisymmetric no angular momentum integral exists. Numerical orbit calculations indicate that effective integrals in addition to E do exist, but it is doubtful that these integrals can be written down explicitly. Thus again—with the exception of some very special cases—it may well be that numerical calculations on the evolution of such systems may provide the only adequate way to deal with them. Before leaving this subject we should point out one common misconception. It is frequently stated that the bar-like structure can only persist if the system is in a state of solid body rotation. Clearly it is true that this must apply to the equi-density contours, but this does not imply that the stellar velocity field is one of solid rotation. In fact, Prendergast has studied barred spirals with considerable internal circulation, very different from solid rotation. And also observational evidence shows that solid rotation of even the gas in these systems is the exception rather than the rule.

IId. *Moments of the Boltzmann equation.* In the preceding discussion we have already discovered that an exact analysis of the Boltzmann equation is quite difficult. Also we found that more semiempirical methods can give us quite useful results. In the present section we shall deal with the moments of the Boltzmann equation, which have proved so useful in hydrodynamics. With the help of some of the results of the previous discussion and with some empirical data we will derive in a systematic way a number of useful results, which are of course all well known, but which have sometimes been derived on shakier assumptions than is needed.

The moment equations are obtained by multiplying the Boltzmann equation with various powers and products of velocity components and then integrating the resulting equation over velocity space. Let us define the number density n, the mean macroscopic velocity $\mathbf{u}$ and the pressure tensor P_{ij} by the following relations:

$$n = \int f d\mathbf{v},$$

(37) $$n u_i = \int f v_i d\mathbf{v},$$

$$P_{ij} = m \int f(v_i - u_i)(v_j - u_j)\, d\mathbf{v} = m \int f w_i w_j d\mathbf{v},$$

where $\mathbf{v}$ is the velocity of a particle and $\mathbf{w} = \mathbf{v} - \mathbf{u}$ is the random part of the velocity. Clearly P_{ij} (the j-momentum transferred per unit time in the i-direction) is a symmetric tensor: $P_{ij} = P_{ji}$. If $P_{ij} = 0$ for $i \neq j$ the pressure tensor is said to be diagonal, while if $P_{ij} = P \delta_{ij}$, that is, when the three diagonal elements are equal, the pressure is a scalar. Of course we could continue to introduce tensors of higher rank, like the heat flow tensor τ_{ijk}, etc.

To obtain the zeroth-order moment equation we integrate the Boltzmann equation over velocity space. Since the x_i, v_i and t are independent variables we have

(38) $$\frac{\partial}{\partial t} \int f d\mathbf{v} + \frac{\partial}{\partial x_i} \int f v_i d\mathbf{v} - \frac{\partial \Phi}{\partial x_i} \int \frac{\partial f}{\partial v_i} d\mathbf{v} = 0.$$

Integrating the third term by parts we see that it vanishes, since at the limits of integration ($\pm \infty$) f vanishes. Inserting the macroscopic quantities from Equation (37) we obtain the continuity equation

(39) $$\partial n / \partial t + \partial (n u_i) / \partial x_i = 0.$$

Of course it could have been derived more directly from simple geometrical considerations. Next we consider the first-order moments. We multiply the Boltzmann equation with $m v_j$ and integrate:

(40) $\quad m\dfrac{\partial}{\partial t}\displaystyle\int fv_j d\mathbf{v} + m\dfrac{\partial}{\partial x_i}\displaystyle\int fv_iv_j d\mathbf{v} - m\dfrac{\partial\Phi}{\partial x_i}\displaystyle\int v_j\dfrac{\partial f}{\partial v_i}\,d\mathbf{v} = 0.$

After integration by parts the third integral is transformed into $-n\delta_{ij}$. The second integral becomes

(41) $\quad \displaystyle\int fv_iv_j d\mathbf{v} = \int f(u_iu_j + u_iw_j + u_jw_i + w_iw_j)\,d\mathbf{v} = nu_iu_j + \dfrac{1}{m}P_{ij},$

since integrals like $\int fw_i d\mathbf{v}$ obviously vanish. Thus we have

(42)
$$mn\dfrac{\partial u_j}{\partial t} + mu_j\left[\dfrac{\partial n}{\partial t} + \dfrac{\partial}{\partial x_i}(nu_i)\right]$$
$$+ mnu_i\dfrac{\partial u_j}{\partial x_i} + \dfrac{\partial P_{ij}}{\partial x_i} + mn\dfrac{\partial\Phi}{\partial x_j} = 0.$$

The expression between brackets vanishes on account of the continuity equation. Thus in vector notation we have

(43) $\qquad \rho\left[\partial\mathbf{u}/\partial t + (\mathbf{u}\cdot\nabla)\mathbf{u}\right] = -\nabla\cdot\mathbf{P} - \rho\nabla\Phi,$

quite similar to the usual hydrodynamic equation of motion, except for the fact that the gradient of the scalar pressure is replaced by the divergence of the pressure tensor. In conventional hydrodynamics it would now be assumed that collisions—which do not yet enter explicitly in the first order momentum equations—would make P_{ij} isotropic. Then it could further be assumed that processes would be adiabatic—leading to $P = f(\rho)$; or alternatively the ideal gas law ($P = R\rho T/\mu$) could be assumed with the temperature T determined from some macroscopic law of heat transport. In both cases we have a closed set of equations.

But if collisions are absent such a procedure is not justified. Then the first moment equations involve the second-order moments (P_{ij}), the second moment equations involve the third-order moments, and so on. Thus we have always one more variable than equations, the moment equations are an infinite set, and it might well be simpler to analyze the Boltzmann equation directly. Still the low-order moment equations can be useful provided they are supplemented by some plausible assumptions or empirical evidence.

IIe. *Moment equations applied to the Galaxy.* Let us consider a galaxy which is steady and axisymmetric ($\partial/\partial t = 0$, $\partial/\partial\phi = 0$). Writing Equation (11) in velocity variables and dropping the ϕ and t-derivatives we obtain the Boltzmann equation applicable to this case,

$$(44) \quad v_{\varpi} \frac{\partial f}{\partial \varpi} + v_z \frac{\partial f}{\partial z} + \frac{1}{\varpi} (v_\phi^2 - \Theta^2) \frac{\partial f}{\partial v_{\varpi}} - \frac{v_{\varpi} v_\phi}{\varpi} \frac{\partial f}{\partial v_\phi} - \frac{\partial \Phi}{\partial z} \frac{\partial f}{\partial v_z} = 0.$$

Here the ϖ-derivative of the gravitational potential has been replaced by the circular velocity (the velocity a particle would have when moving in a purely circular orbit) by Equation (35).

Since f is a solution of the Boltzmann equation we would have $f = f(E, J_z, I_3)$. If we consider a flat galaxy like our own then for low velocity stars a separable potential may be assumed, and I_3 is given by Equation (30). Then f is quadratic in v_{ϖ} and v_z, and all moments odd in v_{ϖ} or v_z vanish. In particular, therefore, we should have $u_{\varpi} = 0$ and $u_z = 0$. For stars of higher velocity the potential can no longer be regarded as separable, but we shall extrapolate our assumption that f is even in v_{ϖ} and v_z. This also seems reasonable in view of the lack of skewness in the observed v_{ϖ} and v_z distributions near the sun.

Clearly f cannot be even in v_ϕ since the system is rotating, but it has frequently been assumed that f is even in w_ϕ. This assumption is not compatible with a finite velocity of escape for the system. In a steady state no stars should be present with velocities in excess of the velocity of escape. Thus:

$$(45) \qquad v_{\varpi}^2 + v_z^2 + u_\phi^2 + 2u_\phi w_\phi + w_\phi^2 < V_e^2,$$

and the distribution in w_ϕ must be skew as it is actually observed to be for large velocities. Condition (45) also shows that the distribution function may not be factorized into functions depending on v_{ϖ}, v_z or w_ϕ only.

In the following we shall write the moments as bracketed quantities, and we shall write the moments in terms of the random velocity components v_{ϖ}, w_ϕ and v_z. Thus

$$\langle v_{\varpi}^\alpha w_\phi^\beta v_z^\gamma \rangle = n^{-1} \int f v_{\varpi}^\alpha w_\phi^\beta v_z^\gamma d\mathbf{v},$$

and if α or γ are odd the moment vanishes.

Of the first-order moment equations only those obtained by multiplying the Boltzmann equation with v_{ϖ} and v_z contain non-vanishing terms. We obtain

$$(46) \qquad \partial (\varpi n \langle v_{\varpi}^2 \rangle)/\partial \varpi + n(\Theta^2 - u_\phi^2) - n \langle w_\phi^2 \rangle = 0$$

and

(47) $$\partial (n \langle v_z^2 \rangle)/\partial z + n \, \partial\Phi/\partial z = 0.$$

The first equation can be written as

(48) $$\Theta^2 - u_\phi^2 = \langle w_\phi^2 \rangle - (1/n) \, \partial(\varpi n \langle v_\varpi^2 \rangle)/\partial \varpi$$

and thus expresses the difference between the rotational velocity and the circular velocity in two terms which contain the velocity dispersions. These terms are readily understood. If the random motions are high the orbits of the stars are highly elliptical. Near its apocenter a star in an elliptical orbit has a smaller azimuthal motion than a star that moves in a circular orbit through the same point; if at that point the star would be near its pericenter the reverse is true. Thus if at a given point there are more stars in the apocenter of their orbits than at the pericenter, the stars have a root mean square azimuthal motion $(\langle v_\phi^2 \rangle)^{1/2} = (u_\phi^2 + \langle w_\phi^2 \rangle)^{1/2}$ that is less than the circular velocity. And this is the case if the stellar density or the magnitude of random motions have a large negative gradient. In discussing stellar populations we will note that stars with large random motions and high density gradients have rather low rotational motions around the galactic center. This fact thus is easily understood on the basis of Equation (48). Also by making use of this equation we can in principle obtain Θ from the observed motions of the stars.

If in Equation (47) we take $\langle v_z^2 \rangle$ constant we recover Equation (34), which we have already made use of to obtain the mass density near the sun.

Under the conditions that have been stated before only two second-order moment equations yield new information; the other four contain only moments that are odd in v_ϖ or v_z. Multiplication of the Boltzmann equation with $v_\varpi w_\phi$ and with $v_z w_\phi$ leads after integration to the equations

(49) $$(1/\varpi) \, \partial \left[\varpi^2 n (u_\phi \langle v_\varpi^2 \rangle + \langle v_\varpi^2 w_\phi \rangle) \right]/\partial \varpi$$
$$+ n u_\phi (\Theta^2 - u_\phi^2 - \langle w_\phi^2 \rangle) - 2 n u_\phi \langle w_\phi^2 \rangle - n \langle w_\phi^3 \rangle = 0,$$

and

(50) $$\partial \left[n (u_\phi \langle v_z^2 \rangle + \langle w_\phi v_z^2 \rangle) \right]/\partial z + n u_\phi \, \partial\Phi/\partial z = 0.$$

Eliminating in the first equation the term involving u_ϕ by making use of Equation (48) and in the second equation by Equation (47) we obtain after some simple reductions

$$(51) \qquad \langle v_\varpi^2 \rangle \frac{\partial (\varpi u_\phi)}{\partial \varpi} = 2 \langle w_\phi^2 \rangle u_\phi + \langle w_\phi^3 \rangle - \frac{1}{\varpi n} \frac{\partial}{\partial \varpi} (\varpi^2 n \langle v_\varpi^2 w_\phi \rangle)$$

and

$$(52) \qquad \langle v_z^2 \rangle \frac{\partial u_\phi}{\partial z} = -\frac{1}{n} \frac{\partial}{\partial z} (\langle w_\phi v_z^2 \rangle n).$$

If the random velocities are small compared to u_ϕ the last two terms in Equation (51) are quite small, especially if the w_ϕ-distribution is not too skew, and we find for the ratio of radial and azimuthal velocity dispersions

$$(53) \qquad \frac{\langle w_\phi^2 \rangle}{\langle v_\varpi^2 \rangle} \cong \frac{1}{2 u_\phi} \frac{\partial (\varpi u_\phi)}{\partial \varpi} = \frac{1}{2} \frac{\partial \ln (\varpi u_\phi)}{\partial \ln \varpi}.$$

Here the effects of the absence of collisions become noticeable. A collision term would tend to make the two dispersions equal. If $u_\phi \propto \varpi^{-1}$, $\langle w_\phi^2 \rangle$ would vanish according to this relation. This is consistent with the fact that a star in an elongated orbit would change its v_ϕ also as ϖ^{-1} because of conservation of angular momentum.

Equation (53) actually provides a comparatively reliable way to obtain $\partial (\varpi u_\phi) / \partial \varpi$ from the observed velocity dispersions. Equation (52) shows that the neglect of the moments odd in w_ϕ indeed leads to doubtful results, as $\partial u_\phi / \partial z = 0$ is unlikely in a finite system. In Appendix II we briefly discuss the third-order moments.

For low velocity stars in our Galaxy the results obtained from the Boltzmann equation can also be found from a discussion of stellar orbits. Since, in some problems, the orbit theory can be very convenient, it is summarily discussed in Appendix III.

III. **Galaxies: contents and motions.** In most galaxies—like in our own—stars make up the bulk of the matter. Because of their long relaxation times they move effectively as free particles in the gravitational field of the galaxy. The interstellar medium is a quite different galactic constituent. Included in it are the interstellar gas, dust, magnetic fields and cosmic rays. The mean free path in the gas is short compared to characteristic lengths in the system, and hydrodynamics may frequently be applicable. Because it is a rather effective radiator, the gas is a highly dissipative medium. The dust, which has a negligible mass, has no direct dynamical effects, but it may affect the composition and energy balance of the gas. The magnetic fields and the cosmic rays may be of some dynamic importance.

Gas and stars are not independent components of a galaxy. The stars form continuously from the gas. At the end of their evolution, the more massive stars reinject most of their matter into the gas. The nuclear reactions inside the stars provide the energy for many processes in the interstellar medium and change its composition. The gas and the stars collectively produce the gravitational field in which both move.

The stars differ in mass ($< 0.1 \odot - 100 \odot$) and thus, in time rate of evolution. Massive stars evolve on time scales of millions of years; stars of the solar mass need several billions of years. Thus, the massive stars are seen in close association with the gas from which they formed, but stars of low mass can give us information on conditions in the galaxy long ago. In most galaxies, stars of small mass appear to contribute most to the mass, but since the luminosity of a star is proportional to the cube or fourth power of its mass, the very massive stars contribute an important fraction of the total light. Thus, the light distribution in a galaxy need not always be representative for the mass distribution.

Detailed observations of the stars and the gas can be made only in our own galaxy and for many objects only in the immediate neighborhood of the sun. Thus, a large part of our discussion will be concerned with our galaxy, but whenever possible, comparison with other galaxies should be made. We shall first consider the contents and the structure of galaxies. Subsequently we shall review the kinematics of stars and gas in our galaxy. Where possible, we shall relate the observed situation to the theoretical discussions of the previous lectures.

IIIa. *Stellar populations.* A theorem on stellar structure states that for a given chemical composition the structure of a chemically homogeneous, static and nonmagnetic star depends on the mass only. If a star evolves, helium or heavier elements become more abundant in its interior, and the star will move along a certain track in the Hertzsprung-Russell diagram. This track again depends only on the mass and initial composition. Thus a star—aside from perturbing effects due to rotation and magnetic fields—can be characterized by three parameters: mass, age and initial (age zero) composition. The stars have condensed out of an interstellar medium that gradually has acquired appreciable abundances of heavier elements. Originally the gas may have been pure hydrogen (plus possibly helium), but the stars have injected elements into the gas which had been synthesized in their interiors. Thus a correlation between age and initial composition of a star might be expected. Such a correlation appears indeed to exist, but fluctuations may well be large. If the correlation were good enough, two parameters, mass and age, would characterize a star completely. There is evidence that in the first billion years of our galaxy, most stars were formed and also most of the heavy elements synthesized. Since for older stars age determinations have uncertainties well in excess of 10 percent, chemical composition may be a more suitable parameter than age for relating older stars to stages of galactic evolution.

Next we look for a moment at the kinematical properties of stars. For stars that are not too distant from the sun, individual space velocities can be determined. The velocities of the stars can be decomposed in a mean velocity for the whole group of stars and individual random motions around the mean. The main systematic motion in our galaxy is one of rotation around the galactic center. If now groups of stars of different ages are analyzed, it appears that for older stars the velocity dispersion is larger and the systematic rotational velocity is smaller than for younger stars. The older stars reach greater heights above the galactic plane, because of their larger z-velocities, and also appear to have a much stronger density gradient towards the galactic center than the young stars. Thus it seems that the age of the stars has a direct correlation to their composition, kinematical characteristics and spatial distribution.

We are now in a position to appreciate the notion of a stellar population, (sometimes referred to as a stellar subsystem), a fruitful concept first introduced in a systematic way by W. Baade. Somewhat loosely, a stellar population is defined as a group of stars which may differ in mass, but which have the same age, composition, kinematical characteristics and spatial distribution. Clearly the last two characteristics are of a statistical nature and thus individual stars should be assigned to a population on the basis of age and composition or some related characteristic, like a conspicuous type of light variability. It is fruitful to introduce the populations just because they behave kinematically differently. Thus in the dynamical analysis we should consider the gas of stars to be a multicomponent gas, and we have a criterion (age and composition) to assign a star to a certain component.

Originally Baade introduced two populations only. Population I contained the interstellar gas and most of the stars found in the solar neighborhood. Population II was composed of the stars found in the halo and central regions of the galaxy (Figure 1) like globular clusters and RR Lyrae stars. The populations were primarily characterized by their Hertzsprung-Russell diagrams (Figure 2).

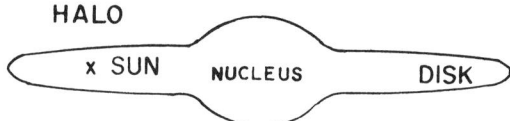

FIGURE 1. Sketch of our galaxy

In population I the stars along the main sequence are unevolved stars which differ in mass, while the giants are stars in advanced evolutionary stages, and the white dwarfs are the remnants of stars that have completed their evolution. In the old population II the massive stars have all long ago completed their evolution and the main sequence extends only over the lower part of the diagram. After a star leaves the main sequence, the subsequent evolution proceeds fast and the star moves up the giant branch and then traverses the horizontal branch of the diagram. Certain easily recognizable variable stars were shown to belong to specific populations and served as tracers. RR Lyrae type stars (periodic variables with periods around 0.5 day) are found only in population II, type I cepheids only in population I. The same populations were also recognized in other galaxies. In elliptical galaxies where not much gas, dust or bright stars are in evidence, population II was supposed (not quite correctly) dominant, and the same applies to the nuclear regions of spiral galaxies. But in the disks of spiral galaxies and in most irregular systems, gas and dust are much in evidence, and thus population I is the main constituent of these regions.

Gradually it has become evident that populations I and II

TABLE III. Populations in our galaxy

	Extreme II	Intermediate II	Disk	Older I	Extreme I
Population	Extreme II	Intermediate II	Disk	Older I	Extreme I
Subsystem	Spherical	Intermediate	Disk	Flat	Very flat
Typical members	Globular clusters RR Lyrae stars	Long period variables	Weak line stars Planetary nebulae	Strong line stars A stars	Gas O-B stars
$\langle z \rangle$ (pc)*	2000	700	400	160	120
$-\partial \log \rho / \partial \log \varpi$ †	4	2	1	0	0
Rotational velocity (km/sec)	< 100	200	240	250	250
Velocity dispersion (km/sec)	250	80	25	15	12
Age (10^9 years)	$10^{††}$	$10^{††}$	$2-10^{††}$	$0.2-2$	< 0.2
Abundance of heavy elements	0.001	0.01	0.02	0.03	0.04
Fraction of galactic mass	0.2 (?)	0.2 (?)	0.5 (?)	0.1	0.03

* $\langle z \rangle$ is the height above the galactic plane where the density has fallen by a factor of 2.

† $\partial \log \rho / \partial \log \varpi$ is the logarithmic density gradient in the galactic plane.

†† The age of the older stars in our galaxy still is uncertain by at least 50 percent. But if a change is made in one of these figures, the two others should be changed also, by approximately the same factor.

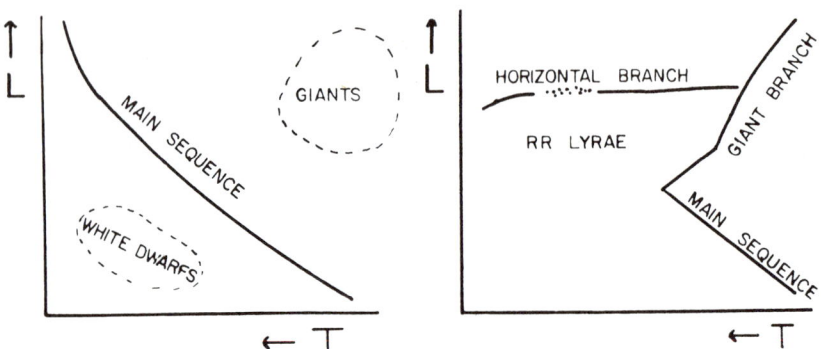

FIGURE 2. Schematic Hertzsprung-Russell diagram
for Baade's populations I and II. The horizontal scale is about
linear in the inverse of effective surface temperature (hotter
stars to the left). Usually an observational parameter, like
spectral type or color index $(B - V)$ is given, with $(B - V)$
equal to minus 2.5 times the logarithm of the ratio of intensities
in specified wavelength intervals in the blue and the yellow.
Along the vertical scale the luminosity is given, usually
expressed in the visual absolute magnitude, which is minus
2.5 times the logarithm of the visual flux plus a constant.
The sun is at $M_v = 4.6$.

included a wide variety of objects and that more subdivisions are
needed. In reality there is a continuous transition from extreme
population II to extreme population I objects. Nevertheless, the
division in some discrete groups remains useful, but care is needed
to define the groups in a sufficiently precise manner. In Table III
the properties of five populations, in which the stars in our galaxy
may conveniently be divided, are given. Numerical values are
indicative only. The table is largely based on a system introduced
by Oort. The term subsystem is frequently used as equivalent to
population. It may be advantageous to use it more explicitly when
the common kinematical characteristics of groups of stars are
stressed.

The population concept is so useful because of the correlation
between composition or age and kinematical characteristics. This
correlation can be understood on the basis of present ideas on
galactic evolution. The system probably started as a gas cloud with
high random motions. These motions dissipated gradually and as
a consequence the system flattened, since contraction towards

the rotation axis was largely prevented by angular momentum conservation. Stars formed in the first phase constitute a more or less spherical subsystem. Gradually flatter subsystems were formed until the gaseous disk reached its present thinness. At the same time, nuclear evolution of the gas took place. Stars that had been formed in the first phase ejected gas enriched in heavy elements. Thus the kinematical and nuclear evolutions are coincidental, and the correlation of these parameters with age is understandable. The correlation between rotational velocity, velocity dispersion, and density gradient is consistent with the results from the first-order moments of the Boltzmann equation. The same is true for the relation between $\langle z \rangle$ and the velocity dispersion.

IIIb. *Types of galaxies.* Various morphological classification schemes for galaxies have been proposed from time to time, but only Hubble's original system has won wide acceptance. Ultimately when more spectroscopic and detailed photometric data are available, one may well feel the need to make refinements, but it would seem that these could find their place within Hubble's system. The Hubble classification—with some modifications due to Sandage—divides galaxies into four broad groups: elliptical and S0 galaxies, spirals, barred spirals and irregulars.

1. *Elliptical and* S0 *galaxies.* The light distribution in these galaxies is smooth. Signs of dust are rather rarely seen and no gas is in evidence, except for an occasional small emission nucleus. The integrated spectra do not indicate large heavy element deficiencies, and the dominant population apparently is of disk type. In M 32, the dwarf elliptical companion of the Andromeda nebula, some population II stars are also seen.

The ellipticals are divided in eight groups: E0—E7. The isophotes are represented extremely well by ellipses. A galaxy is classified En if the axial ratio satisfies $n = 10(1 - b/a)$. The ellipticals appear to be axisymmetric oblate spheroids. It is clear that the seemingly circular E0 galaxies may contain highly oblate objects seen pole on. According to Sandage, the observed frequency distribution of ellipticals is consistent with the assumption that the abundance of the different classes would be about equal, if all were seen edge on.

Ellipticals flatter than E7 have not been found. Flatter systems always possess a disk. They are classified S0 systems and form the transition type between ellipticals and spirals. Spiral structure, however, is completely absent in S0 systems. In some clusters of galaxies, S0 systems are abundant and normal spirals rare. Baade and Spitzer have suggested that these S0's are spirals that have lost their gas in collisions; if galaxies collide, the stellar systems would pass through each other without hindrance, but the gas of both systems might be left behind. It seems doubtful that all S0 galaxies can be explained along these lines.

2. *The spirals.* Structural features, gas, and dust become visible in the spirals. The systems consist of a disk in which spiral structure is embedded and a central concentration which, in the absence of the disk, would not look very different from an elliptical. The spiral sequence runs from S*a* through S*b* to S*c*. In an S*a* the nucleus is large and bright, the spiral arms are thin and would tightly around the nucleus. In the S*c*'s, the nucleus is inconspicuous; massive, open, spiral arms dominate the picture. The nuclei probably have the same composition as the ellipticals. The disks contain mainly disk population and population I. The extreme population I, the gas and the bright stars appear to be confined to the neighborhood of the spiral arms. It is not known what fraction of the mass of the disk is in the arms, but it is unlikely to be more than 10 percent. Thus the mass distribution, and therefore the gravitational field, are probably basically axisymmetric.

3. *The barred spirals.* Axial symmetry is obviously absent in the case of the barred spirals. In these objects, the basic feature is a bar in the middle. At the ends of the bar the spiral arms appear. The bar presumably is something like a prolate spheroid. The barred spirals form a sequence parallel to the ordinary spirals SB*a*, SB*b* and SB*c*. Also SB0 systems occur in which the arms are frequently replaced by a ring with the bar like a spoke in a wheel. The barred spirals may be somewhat less abundant than the ordinary spirals, but the difference is not very large. No clear systematic differences in mass, luminosity, and size between the two groups have been established, but the observational material is quite limited.

4. *The irregulars.* In the late S*c*'s, the arms are massive and irregular. Gas and dust are dominant, but the systems still have large scale symmetry. In the irregulars this is no longer the case. In type I irregulars gas and dust are much in evidence, as in the large Magellanic Cloud. But type II irregulars contain mainly

population II stars. In this latter group belong the Sculptor systems, very loose aggregates of stars without interstellar matter. Masses and luminosities are low and they are difficult to detect. In the local group of galaxies to which our galaxy and the Andromeda—both Sb systems—belong, the number of Sculptor systems is larger than that of all other types together, but their contribution to the total mass is negligible. Only three percent of the brighter galaxies belong to the class of the irregulars, but as many are intrinsically faint, they are not rare.

Frequently the question has been raised whether the sequences E-S0-Sa-Sb-Sc-Irr or E-SB0-SBa-SBb-SBc-Irr have evolutionary significance. It was suggested some time ago that galaxies began their life as ellipticals, which then because of some rotational instability developed a disk and spiral arms. Later when the importance of the gas in the spiral structure was recognized, the inverse sequence was suggested. A galaxy would start out as an irregular turbulent gas cloud and as the turbulence diminishes it will become more and more regular. Since the conversion of gas into stars is ultimately an irreversible process, such a sequence through stages of lesser gas content seems plausible, and it is not unlikely that an individual galaxy may have indeed at its origin looked like an irregular, and late in its life like an S0. It appears doubtful, however, whether a spiral or an S0 can evolve within the available time scale into an elliptical. The ellipticals appear to have comparatively large random motions and low angular momentum. Thus an elliptical may well have originated directly from a more irregular system. However, the question as to whether the galaxies we see nowadays are different evolutionary phases of effectively the same system must probably be answered negatively. In Table IV some data on galaxies are given, and there is a trend for the ellipticals and early spirals to be more massive than the later spirals and irregulars. Thus, if mass is conserved—and even in the presence of a suitable intergalactic medium it is doubtful whether a significant amount of gas could accrete in the available time—the types which are now observed must have had different initial conditions.

A related question is whether all galaxies originated about at the same time, or whether galaxies are also forming presently. No definite conclusion has been reached as yet, but there are certainly galaxies and groups of galaxies that have the appearance of being young. Of course these could also be older objects passing through a rapid evolutionary stage. On the other hand it seems that most galaxies in our immediate neighborhood are of comparable age. It should be stressed, however, that whatever one's cosmological prejudice, both situations could be envisaged. In a steady state cosmology, it has been suggested that galaxy formation takes place simultaneously over rather large regions, while in an exploding cosmology, in certain regions enough gas might be left to form new galaxies.

In Table IV we have assembled some data on the various types of galaxies. The first column gives the abundance of the type among the brighter galaxies, the second the estimated abundance per unit volume determined from the mean luminosity in units of the sun's luminosity given in the third column. Then follow the mass to light ratio expressed in solar units, the mean mass in solar masses, and finally the ratio of the mass of observable gas (mainly neutral and ionized H) to the total mass. Some of the numbers are highly uncertain. The masses can be obtained in principle from observations of multiple galaxies. For spirals the rotational motions of the gas enable us to obtain fairly good masses and sometimes density distributions. The relationship between mass models of galaxies and the rotational velocities is further discussed in Appendix IV.

TABLE IV. Properties of Galaxies

Type	Relative Frequency		$L(10^{10}\odot)$	M/L	$M(10^{10}\odot)$	$M\,\mathrm{gas}/M$
	Among Bright Galaxies	In Unit Vol.				
E	12	6	0.2	50	10	$[10^{-6}]$
S0	11	6				
Sa	8	4				
Sb	33	17	0.2	20	4	0.05
Sc	33	49	0.1	10	1	0.1
I	3	[18]	0.04	5	0.2	0.2

Data on galaxies other than our own are so scarce and uncertain that a very direct confrontation between the theory developed before and the observations seems hardly possible. It would appear most fruitful therefore to construct complete theoretical models for a wide variety of galaxies and to see whether some such models can account for the light distribution and perhaps some other properties of the galaxies. At the same time such models would be helpful in deriving masses from very fragmentary kinematical data. In galaxies where gas is abundant, the situation is further complicated by the uncertain gas dynamic effects.

IIIc. *Stellar motions in the solar neighborhood*. Somewhat complete kinematical data can be obtained for the gas only in very nearby galaxies and for the stars only in our own system. We thus restrict the further discussion to our galaxy. Some qualitative data have been given already when we discussed stellar populations, but now we turn to a more precise and systematic discussion.

For the nearby stars (within a few hundred pc from the sun) good space velocities can be obtained from radial velocities and proper motions. At larger distances, the proper motions usually are too small and only the radial velocity component can be obtained. We again denote the velocity of a star by v, the mean velocity of a group of stars by u, and the random component of velocity by w.

Our first task is the determination of the mean motion u for the various populations near the sun. The absolute determination of u is difficult, but relative values of u can be obtained by measuring the mean motion of the sun with respect to different groups of stars. The solar motion is $v(\odot) - u$ and the difference of the solar motion with respect to two groups of stars is equal to the difference of $- u$ for the two groups. Solar motions determined for some populations are given in Table V, where the root mean square random velocities are also listed.

Inspection of the table shows that no significant differences are found in u_ϖ and u_z for different groups, and this suggests that rotation around the galactic center is the only large scale motion of the stars near the sun. This will be confirmed by our later analysis of the local velocity field. However, there are significant differences in the mean rotational velocities of the stars. Since the entries in the table are arranged more or less in an age sequence, we see that the older stars rotate more slowly than the younger stars. For the real population II objects like RR Lyrae stars and globular clusters, the effect is very conspicuous. The random motions are becoming clearly larger as the rotational velocity decreases and this is, of course, to be expected from Equation (48).

The most important feature of the distribution function of random velocities is its anisotropy. The dispersion in the z-direction is smallest, but also in the plane the function is not isotropic, with the radial motion the largest. The latter situation is predicted by Equation (53). Inserting the numerical values for the rotation parameters given in Equation (64), this equation appears in fact satisfied as well as can be expected. The velocity dispersions in the table were derived by assuming the Schwarzschild ellipsoidal distribution of random velocities, that is

$$(54) \qquad f(\mathbf{w}) \propto \exp(- \alpha^2 w_1^2 - \beta^2 w_2^2 - \gamma^2 w_3^2)$$

with w_1, w_2, w_3 orthogonal velocity components. One of the axes of the velocity ellipsoid coincides with the z-direction, but in the plane the direction of maximum velocity dispersion (vertex, direction of star streaming) may make an angle of the order of 10° or more with the direction to the galactic center, at least for the younger stars. This is probably due to local irregularities.

We have already pointed out that the Schwarzschild distribution is incompatible with a finite velocity of escape for the galaxy. Also, from the observational point of view, the Schwarzschild distribution is clearly not valid for large velocities. Oort has shown, more than 30 years ago, that the stellar residual velocities in excess of 65 km/sec show a very marked asymmetry. Hardly any stars with such velocities in the direction of galactic rotation ($l = 90°$) are found, but there are stars with these velocities in the opposite direction. It is clear, therefore, that a distribution function which is even in the random velocities cannot fully represent the data.

Let us now consider the systematic velocity field around the sun. We suppose that the velocity field is sufficiently smooth for a Taylor expansion to be permissible. Thus we write

$$(55) \qquad \mathbf{u}(\mathbf{r}) = \mathbf{u}(\odot) + (\mathbf{r} \cdot \nabla) \mathbf{u} + \cdots$$

where $\mathbf{r}$ is the radius vector from the sun to the point where the velocity is considered. We consider stars that are situated in the plane of the galaxy and we take $u_z = 0$. Here and in the following, if the velocity of the sun is mentioned, it is the systematic mean motion of the younger populations around the sun, with the

TABLE V. Solar motion and random velocities
for different kinds of stars. The average age is
increasing downwards.

Type	$v_\varpi(\odot) - u_\varpi$	$v_\phi(\odot) - u_\phi$	$v_z(\odot) - u_z$	$\langle w_\varpi^2 \rangle^{1/2}$	$\langle w_\phi^2 \rangle^{1/2}$	$\langle w_z^2 \rangle^{1/2}$
B0	11	− 18	8	12	9	4
A0	10	− 11	6	17	9	7
F0	10	− 11	6	23	13	12
dG0	12	− 16	7	32	21	18
dK0	10	− 18	6	39	25	18
dM0	10	− 22	8	45	27	18
RR Lyrae	11	− 127	29			
Glob. clusters, sub dwarfs RR Lyrae	12	− 175	9			

peculiar motion of the sun already taken out. In Cartesian coordinates we have

$$u_x(\mathbf{r}) = u_x(\odot) + x\, \partial u_x/\partial x + y\, \partial u_x/\partial y + \cdots,$$
(56)
$$u_y(\mathbf{r}) = u_y(\odot) + x\, \partial u_y/\partial x + y\, \partial u_y/\partial y + \cdots,$$

where all derivatives are evaluated at the solar position. The radial
and tangential velocities that are measured from the sun, V_r and
V_l are given by (Figure 3)

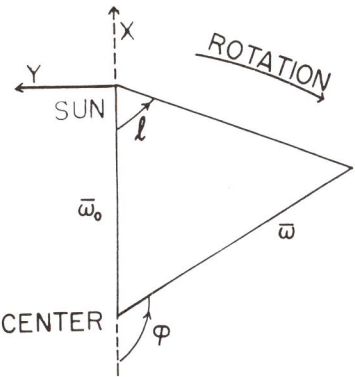

FIGURE 3. Coordinate systems for the description of our galaxy

(57)
$$V_r = - [u_y - u_y(\odot)]\sin l - [u_x - u_x(\odot)]\cos l,$$
$$V_l = - [u_y - u_y(\odot)]\cos l + [u_x - u_x(\odot)]\sin l.$$

Thus, since $x = -r\cos l$ and $y = -r\sin l$,

$$V_r = \frac{r}{2}\left[\left(\frac{\partial u_x}{\partial x} + \frac{\partial u_y}{\partial y}\right) + \left(\frac{\partial u_x}{\partial y} + \frac{\partial u_y}{\partial x}\right)\sin 2l\right.$$
$$\left. + \left(\frac{\partial u_x}{\partial x} - \frac{\partial u_y}{\partial y}\right)\cos 2l\right],$$

(58)

$$V_l = \frac{r}{2}\left[\left(\frac{\partial u_y}{\partial x} - \frac{\partial u_x}{\partial y}\right) - \left(\frac{\partial u_x}{\partial x} - \frac{\partial u_y}{\partial y}\right)\sin 2l\right.$$
$$\left. + \left(\frac{\partial u_x}{\partial y} + \frac{\partial u_y}{\partial x}\right)\cos 2l\right].$$

It is advantageous to transform these expressions into expressions containing only velocity components and derivatives in ϖ and ϕ. Clearly we have

(59)
$$u_x = -u_\varpi\cos\phi + u_\phi\sin\phi, \qquad (\partial/\partial x)_\odot = (\partial/\partial\varpi)_\odot,$$
$$u_y = -u_\varpi\sin\phi - u_\phi\cos\phi, \qquad (\partial/\partial y)_\odot = (1/\varpi_0)(\partial/\partial\phi)_\odot.$$

Hence, we finally obtain

$$V_r = \frac{r}{2}\left\{\left[\frac{1}{\varpi_0}\frac{\partial(\varpi u_\varpi)}{\partial\varpi} + \frac{1}{\varpi_0}\frac{\partial u_\phi}{\partial\phi}\right] + \left[\varpi_0\frac{\partial(u_\phi/\varpi)}{\partial\varpi} + \frac{1}{\varpi_0}\frac{\partial u_\varpi}{\partial\phi}\right]\sin 2l\right.$$
$$\left. + \left[\varpi_0\frac{\partial(u_\varpi/\varpi)}{\partial\varpi} - \frac{1}{\varpi_0}\frac{\partial u_\phi}{\partial\phi}\right]\cos 2l\right\},$$

(60)

$$V_l = \frac{r}{2}\left\{\left[\frac{1}{\varpi_0}\frac{\partial(\varpi u_\phi)}{\partial\varpi} - \frac{1}{\varpi_0}\frac{\partial u_\varpi}{\partial\phi}\right] - \left[\varpi_0\frac{\partial(u_\varpi/\varpi)}{\partial\varpi} - \frac{1}{\varpi_0}\frac{\partial u_\phi}{\partial\phi}\right]\sin 2l\right.$$
$$\left. + \left[\varpi_0\frac{\partial(u_\phi/\varpi)}{\partial\varpi} + \frac{1}{\varpi_0}\frac{\partial u_\varpi}{\partial\phi}\right]\cos 2l\right\},$$

where all velocity derivatives are evaluated at the position of the sun. Thus in the region where the first-order expansion is valid, two results hold independently of the nature of the velocity field:

1. The radial and tangential velocities are proportional to the distance from the sun.

2. Both can be represented as the sum of a constant and a double harmonic wave in galactic longitude.

In the observed radial velocities of stars the coefficient of $\cos 2l$

appears to be essentially zero. The constant term (K-term) is more difficult to determine because systematic errors in radial velocity may produce a pseudo K-term; but probably it is small, if not zero. Thus near the sun $\partial u_\varpi / \partial \varpi = 0$ and $u_\varpi + \partial u_\phi / \partial \phi = 0$. This, and the fact that no differential radial motions between different groups of stars have been noticed, suggests that $u_\varpi = 0$ and that the system is axisymmetric. Then the only motion is the one corresponding to rotation around the center. It should be emphasized that we have no guarantee that this is true throughout the galactic system. If only rotation is present we can write

$$V_r = Ar \sin 2l,$$
(61)
$$V_l = Br + Ar \cos 2l,$$

where the Oort constants of galactic rotation are defined by

$$A = -\frac{1}{2} \left(\frac{u_\phi}{\varpi} - \frac{\partial u_\phi}{2\varpi} \right) = \frac{\varpi}{2} \frac{\partial \Omega}{\partial \varpi},$$
(62)[3]
$$B = \frac{1}{2} \left(\frac{u_\phi}{\varpi} + \frac{\partial u_\phi}{\partial \varpi} \right) = \frac{1}{\partial \varpi} \frac{\partial (\varpi^2 \Omega)}{\partial \varpi},$$

with Ω the angular velocity. To obtain A, we must measure the velocities of stars of accurately known distances. Because of uncertainties in interstellar absorption corrections and absolute magnitudes of the stars, this is not a simple matter, and values between 15 and 20 km/sec/kpc have been suggested. Once A is known, distances of stars can be determined in a statistical manner from their motions. The tangential velocity is proportional to r, and thus the proper motion in longitude is independent of distance in the region where the first-order theory applies. These proper motions contain a constant term related to B, but its determination is extremely difficult because its size is less than $0''.002$/year and difficult to distinguish from systematic errors in proper motions. For the moment, we shall adopt for A, B and ϖ_0 the values

$$A = 15 \text{ km/sec/kpc} = 1.5 \times 10^{-8} y^{-1},$$
(63)
$$B = -10 \text{ km/sec/kpc} = -1.0 \times 10^{-8} y^{-1},$$
$$\varpi_0 = 10 \text{ kpc},$$

---

[3] Because of the way galactic longitudes are counted (Appendix V), the rotational velocity is taken as negative.

corresponding to

$$u_\phi = -250 \, \text{km/sec},$$

$$\partial u_\phi / \partial \varpi = 5 \, \text{km/sec/kpc},$$

where the distance ϖ_0 of the sun to the galactic center also is still somewhat uncertain. It can, in principle, be determined directly by observing the apparent magnitude of the RR Lyrae stars near the galactic center. Since these stars are strongly concentrated near the center, their number as a function of magnitude will show a maximum at the magnitude corresponding to the center. If the absolute magnitude of these stars is known, and if absorption effects can be accurately evaluated, the distance of the sun to the center follows. Our knowledge of these quantities still has considerable uncertainty. Taking all determinations of ϖ_0 together the uncertainty still is of the order of $\pm 2 \, \text{kpc}$. The values in Equation (64) refer to the younger populations near the sun. The random motions of the older populations are so large that the rotation effects cannot be separated out very well. We thus have obtained the rotational velocity and its derivative near the sun. Stellar observations up until now have not enabled us to determine accurate rotational velocities at larger distances from the sun, mainly because interstellar absorption intervenes. Our further knowledge of the galactic rotation curve is therefore based on studies of the 21-cm line emitted by neutral hydrogen.

IIId. *Motions in the interstellar gas.* The interstellar gas consists mainly of hydrogen (70 percent by weight) mixed with helium and also some heavier elements (mainly carbon and oxygen, about 3 percent). Most of the hydrogen is neutral (HI regions), but about a tenth is ionized by ultraviolet radiation from hot stars (HII regions). Atomic hydrogen is observed through the 21-cm line that is emitted when the electron spin switches from antiparallel to parallel with the proton spin. The transition probability is quite low ($3 \times 10^{-15} \text{sec}^{-1}$) and thus the line is intrinsically sharp, but it is broadened by the thermal motions of the hydrogen atoms. Recently lines of the OH radical have also been observed in the radio spectrum. The HI regions can also be studied through the optical absorption lines of Ca^+ and some other impurities. The HII regions can be studied through their Bremsstrahlung (free-free emission) in the radio continuum and through recombination radiation,

---

[4] See footnote 3.

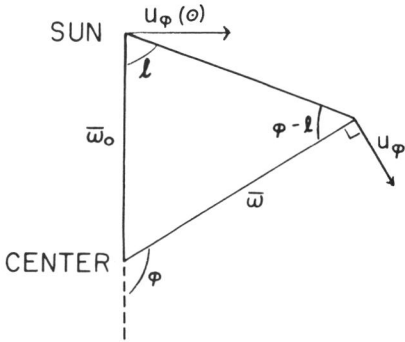

FIGURE 4. Geometry for galactic rotation

especially in the Balmer line H_α. Recently, permitted transitions between very high levels of the H atom ($n \approx 100$) have also been observed in the radio spectrum. Molecular hydrogen could conceivably be even more abundant than atomic hydrogen, but since it cannot be observed in the radio or visible part of the spectrum, nothing is known yet about its presence, and we must wait for the proper satellite observations.

The interstellar gas is concentrated in clouds. These may not be really discrete objects, but the density contrast between the clouds and the intercloud medium appears rather large. Typical clouds have densities of 10 atoms per cm^3, diameters of the order of 10 parsec, masses of a few hundred solar masses, and they fill somewhat less than 10 percent of the total volume. Temperatures in the clouds are typically of the order of $100°$ K in HI regions. Clouds are frequently part of large cloud complexes.

Most clouds are optically thin in the 21-cm line and thus many clouds can be seen along a line of sight. Distances of clouds cannot be obtained, in general. All that can be measured is the number of hydrogen atoms in a certain direction as a function of radial velocity. An exception is the case of optical or 21-cm absorption lines. If one observes such a line one knows at least an upper limit to the distance, since the cloud must lie between the source of the observed radiation and the observer.

Let us suppose that the only systematic motion of the gas is rotation. We consider the radial velocity with respect to the sun of a cloud of gas at longitude l and galactocentric distance ϖ. From Figure 4 we have

(65) $$V_r = u_\phi \sin(l + 180 - \phi) - u_\phi(\odot) \sin l.$$

But

$$(\sin(\phi - l))/\varpi_0 = \sin l/\varpi;$$

hence

(66) $$V_r = \left[\frac{\varpi_0}{\varpi} u_\phi - u_\phi(\odot) \right] \sin l = \varpi_0 [\Omega(\varpi) - \Omega(\odot)] \sin l$$

with Ω the angular velocity.

If $|l| < 90°$ there are two different points along a line of sight with the same value of ϖ and V_r. Except very near the galactic center, Ω is a monotonically decreasing function of ϖ, and along each line of sight V_r reaches a maximum at the point where ϖ has its smallest value. Thus the expected 21-cm profile is as indicated in Figure 5, with a sharp cutoff at V_{max}. Actually the random motions of the clouds remove the sharp features and the actual profiles look more like the second curve in Figure 5. Sometimes it is difficult to decide where in the observed profiles the cutoff is to be located.

From a determination of the maximum velocity as a function of galactic longitude, one thus can find the angular velocity as a

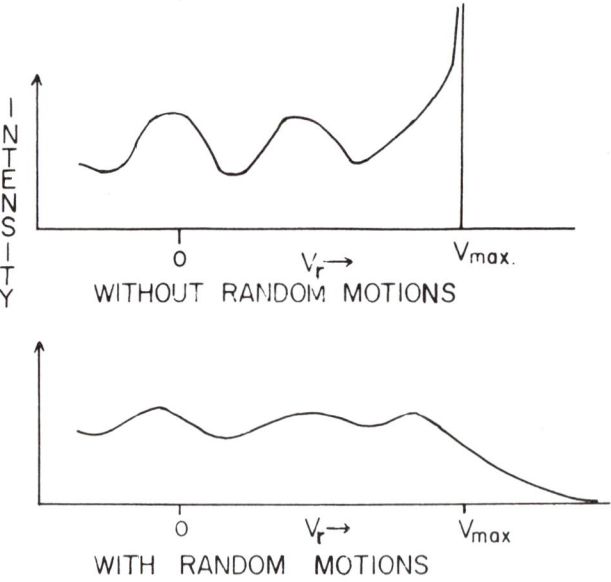

FIGURE 5. Sketch of 21-cm profiles for $|l| < 90°$

function of ϖ between the galactic center and the sun—at least for those regions where the assumption of pure rotation is valid, that is, probably between the galactic center and $\varpi = 700\,\mathrm{pc}$ and between 4 and 10 kpc. Complications arise in practice because of local deviations from purely rotational motions, and because of the nonuniform distribution of the gas. The results that have been obtained are sketched in Figure 6. Between 100 pc and 10,000 pc from the center the linear velocity u_ϕ remains almost constant. It will be seen in Appendix IV that this implies that the density of matter in the galactic plane varies about as ϖ^{-2}. Thus galactic rotation is everywhere very different from solid body rotation.

Some check on the results can be obtained from the fact that the results obtained from observations at $+l$ should agree with those based on observations at $-l$. Differences generally appear to be less than 10 km/sec, but may be somewhat systematic in the sense that negative longitudes tend to give slightly smaller values for $|u_\phi|$. Radio data in the Andromeda nebula also indicate some differences between the two halves of that galaxy. In all further analysis it should be remembered that the rotation curve has been obtained for the *gas*. Little is known about stellar motions far from the sun, but the few data that are available agree with the rotation curve from the gas as well as could be expected. Since these data refer to O and B stars with ages of a few million years only, this is not surprising. Nothing is known directly about the motion of the bulk of the disk stars farther from the sun.

For $|l| > 90°$, V_r has no extremum and is a monotonic function of the distance to the sun. Rotational velocities can only be obtained if use is made of objects of known distance, and not much

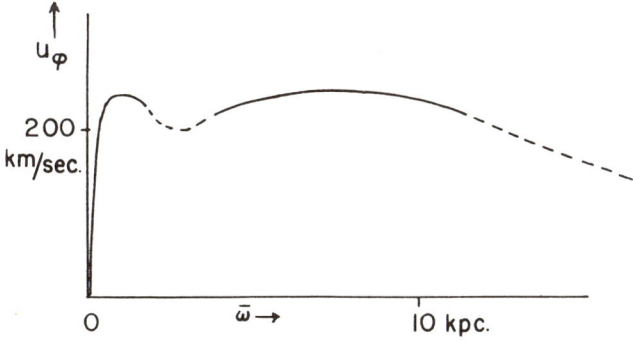

FIGURE 6. Sketch of galactic rotation curve

progress has been made here. The rotation curves commonly used for $\varpi > \varpi_0$ are theoretical extrapolations on the basis of some model for the mass distribution in the galaxy.

The motions of the gas in the immediate neighborhood of the sun also indicate that small deviations from circular motions occur. It seems that matter streams into the galactic plane from both galactic poles, and that in the plane it moves away from us in certain directions. The velocities of these local motions are of the order of 5 km/sec, or a little larger. Recently, however, some clouds have also been observed at high galactic latitude that approach the solar neighborhood with velocities of the order of 100 to 200 km/sec (radial velocity). The nature of these interesting objects is still unclear.

In the central region of the galaxy, the situation is quite different. Large radial motions occur. Because a strong radio source with a continuous spectrum (Sagittarius A) is located at the galactic center, and because the 21-cm line appears in absorption in front of this source, it is possible to separate the gas in front of and behind the center. It appears from this that most radial motions are directed outwards. A conspicuous structure is the "3 kpc arm," a very regular feature with an outward motion of 50 km/sec and a rotational velocity of 200 km/sec. Nearer to the center, large radial motions of up to 150 km/sec are observed. Within 700 pc from the center, however, not much evidence for radial motion is found. Matter appears concentrated in a thin, fast rotating disk. The shear in this disk is large. At 100 pc from the center, the period of rotation is 3×10^6 years, at 200 pc, almost twice as long.

In the quieter regions of the galaxy where the motions in the gas correspond to pure rotation, the distances of elements of gas can be derived from their observed radial velocities, once the rotation curve has been determined. Thus the density distribution in the gas can be found. Results obtained to date show the following. The larger part of the layer of gas is quite flat; within the solar orbit it nowhere deviates more than 50 pc from a plane. But far from the center, the gas layer begins to curve upward in one half of the galaxy and downward in the other half, reaching heights of a thousand pc or more. The thickness of the gas layer is about 280 pc between half density points in the region between the sun and 5 kpc from the center. It decreases to about half of that value further in. Outside the solar orbit it gradually increases. The mean density of atomic hydrogen in the plane near the sun is about 0.5 atom cm^{-3} corresponding to a gas density of 1.2×10^{-24} g cm^{-3}. Inside the solar orbit it does not change much, outside the solar it gradually decreases to very low values around 15 kpc. The total mass of neutral hydrogen is about 3 percent of the mass of the galaxy.

The gas in the disk is concentrated in a number of rather irregular arc-like structures. These arcs may well be part of a spiral pattern. Taking the data at face value the arms seem to be trailing and to be inclined a few degrees with respect to circles around the galactic center. But it may well be that the occurrence of spiral structure is associated with small radial motions. If these were present we would have located the gas incorrectly on the basis of its radial velocity.

IIIe. *Dynamics of the gas, magnetic fields, and spiral structure.* For lack of time we can only give a very brief account of some aspects of the dynamics of the gas. One of the main uncertainties in considerations on the dynamics of the gas in the galaxy is the role of the magnetic field. Information on the strength of the magnetic field comes mainly from the following considerations.

1. Nonthermal radio radiation is observed that originates in the galactic disk and halo. The spectrum and polarization of this radiation indicate that it is synchrotron radiation from relativistic electrons in interstellar magnetic fields. Some information is available on the flux of such electrons near the earth and some reasons can be advanced why this flux should be roughly constant through much of the galaxy. From the flux of electrons and the intensity of the radiation, the strength of the required magnetic field can be inferred. Mean values of about 5×10^{-6} Gauss for the halo and 1.5×10^{-5} Gauss for the disk result.

2. Most of the cosmic rays observed near the earth have probably been confined by galactic magnetic fields for rather long times. It can be argued that effective confinement is only possible if the magnetic field energy density is at least as large as the cosmic ray energy density and this indicates a field of at least 5×10^{-6} G in the halo.

3. The Zeeman effect can be observed in the 21-cm line emitted by HI clouds. Only doubtful positive results have been obtained and the fields indicated are generally less than 5×10^{-6} G. Since

in the Zeeman effect only a large-scale systematic field in a cloud can be detected, the field could be larger if it were nonuniform.

4. The Faraday effect is the rotation of the plane of polarization of a beam of electromagnetic radiation propagating through a medium with magnetic fields and (thermal) electrons. If the wavelength is λ in meters, the electron density $n_e \, \text{cm}^{-3}$, the field $\mathbf{H}$ in μG and the element of length $d\mathbf{l}$ in parsecs, the rotation in radians Ψ is

$$(67) \qquad \Psi = 0.8 \lambda^2 (\text{m}) \int n_e (\text{cm}^{-3}) \, \mathbf{H} (\mu G) \cdot d\mathbf{l}(\text{pc}).$$

Note that Ψ is linear in $\mathbf{H}$, and thus a small-scale random component of the field does not contribute much. The Faraday effect can be observed in polarized extragalactic radio sources, and observations at different wavelengths enable us to find the value of the integral. The observed rotations show a very strong dependence on galactic latitude, and this indicates that much of the effect arises in our galaxy rather than in the sources or in the intergalactic medium. The mean electron density in the galactic disk is still uncertain, but with an upper limit of $0.1 \, \text{cm}^{-3}$ for $\bar{n}_e$ the data indicate typical minimum values of $2 \times 10^{-6} \, \text{G}$ for the mean value of $\mathbf{H}$ along the line of sight. Clearly this could be much lower than the root mean square field.

It is clear from this discussion that in considering the dynamics of the interstellar gas, we have to take into account magnetic fields of the order of $10^{-5} \, \text{G}$, and it is of interest to see how such fields would affect the rotational velocity of the gas. Writing down the virial theorem for a steady closed system of gas, cosmic rays and magnetic fields we have

$$(68) \qquad 2E_{\text{kin}} + E_{\text{c.r.}} + E_{\text{magn}} + E_{\text{grav}} = 0.$$

If we compare now two disks of gas in a gravitational field and if in one of them there are magnetic fields and cosmic rays we will have for the difference in the kinetic energy (if the density distribution is the same)

$$(69) \qquad \Delta E_k = \tfrac{1}{2} \left(E_{\text{c.r.}} + E_{\text{magn}} \right)$$

or, if we take average values over the whole disk and if we suppose only rotational motions are present,

(70) $$\rho u_{\phi} \Delta u_{\phi} \approx -\tfrac{1}{2}\left[e_{\text{c.r.}} + H^2/(8\pi)\right]$$

with $e_{\text{c.r.}}$ the energy density of the cosmic rays. Inserting typical numerical values, $\rho = 1.4 \times 10^{-24} \text{g cm}^{-3}$, $u_{\phi} = 2.5 \times 10^{7} \text{cm/sec}$, $e_{\text{c.r.}} = 1 \times 10^{-12} \text{ergs/cm}^3$ and $H = 1 \times 10^{-5} \text{G}$ we would have $\Delta u_{\phi} = 0.7 \text{ km/sec}$. Of course this is only the mean value for the whole disk, and since the Δu_{ϕ} produced by the field is unlikely to have the same sign everywhere, the local values might be higher. Also, the disk is not a closed system and the magnetic fields of the halo may also be mostly anchored in the disk, in which case they should be included in the total magnetic energy. Nevertheless the above estimate shows that very large differences between the gas velocity and the circular velocity should not be expected, and this provides some justification for making use of the rotation curve in constructing mass models for our galaxy.

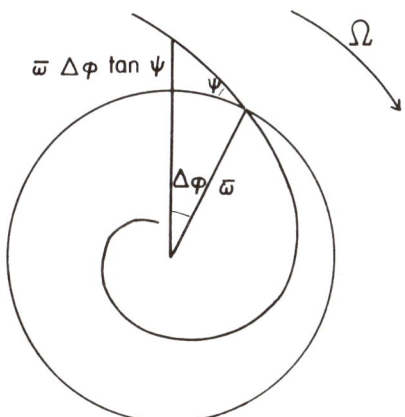

FIGURE 7. Geometry for spiral arm

One of the most conspicuous features of the distribution of the gas is the spiral structure. In our galaxy it is not very easy to decide whether really a large scale spiral pattern is present, but in some other galaxies like M 81 and NGC 5364 this is clearly the case. In a system with differential rotation, the existence of such a spiral pattern that pervades the whole galaxy is hard to understand. For consider a spiral pattern that makes an angle ψ with the azimuthal direction. Consider locally (Figure 7) two nearby points on the spiral arm separated by $\Delta\phi$ in galactocentric longitude. Then the radial distance between these points is $\varpi\Delta\phi\tan\psi$. Let us take the spiral structure to be trailing with respect to the galactic rotation—which is actually the observed situation. Then some time, T_{sp}, in the past the two points were lying along the same radius vector and the spiral arm was locally radial. We have

$$(71) \qquad T_{sp} = \frac{-\Delta\phi}{\varpi\,\Delta\phi\tan\psi\,\partial\Omega/\partial\varpi} = \frac{-T_{rot}}{2\pi\tan\psi\,\partial\ln\Omega/\partial\ln\varpi}.$$

Since in the typical disk region $\Omega \propto \varpi^{-1}$ and since ψ is an angle of $5°$ or $10°$, it is clear that on this picture the spiral pattern should have originated less than a rotation period ago. In view of the frequent occurrence of spiral structure, this appears unlikely.

Two ways out seem possible. It may be supposed that the matter in the spiral arm has also a small radial velocity, so that the velocity vector is everywhere parallel to the arm. Discussion of such a model shows that difficulties arise with the continuity equation, because the radial motion should be partly inward and partly outward, but these difficulties could be resolved if there is much mass exchange with the halo. Also the dynamical situation—which would almost certainly involve hydromagnetic effects—is not without much difficulty.

Alternatively it may be that our implicit assumption that the spiral pattern moves with the same velocity as the matter in the pattern is incorrect. The spiral structure then would be a density wave moving with respect to the gas and the stars. Such density waves probably can be caused by gravitational instability. Since Dr. Lin will extensively discuss this possibility, we shall leave this subject here.

APPENDIX I. Relaxation times

We shall consider a star—the test particle—that moves through a region in which there are many other stars—the field particles— which pass at different distances. We assume that there is no systematic gravitational field, and thus in the absence of the local field stars the test star would move at constant speed along a straight trajectory. We shall find that virtually all encounters between the test particle and the field particles result only in very small deviations from the unperturbed test particle orbit. But the cumulative effect of many encounters can become significant after a long enough time. To formulate the problem more quantitatively let us ask the following questions:

1. How long is the time T_D in which an average test particle is deflected by an angle $\pi/2$ from its original trajectory?

2. How long is the time in which the cumulative energy exchange of test particle and field particles is of the order of the original energy of the test particle? This leads to relaxation times T_{eq} and T_E defined later.

3. How long is the time T_f in which the velocity component of the test particle in the direction of its unperturbed trajectory has decreased on the average by an amount of the order of the original velocity? The retardation of the test particle in the direction of its unperturbed trajectory is referred to as dynamical friction. It is clear that these times can be different for different test particles depending on the particular configuration of field stars along the trajectory. Therefore we shall always consider an ensemble of test particles and take the mean over the ensemble. The time scales we have introduced are not necessarily equal. For example, if the mass of the test particle vanishes and the field particles have no random motions the test particle moves in a constant potential field, and the energy per unit mass then is conserved. Thus the time scales for energy exchange become infinite, but T_D and T_f are finite.

Let us then consider a test particle that encounters a field particle (Figure AI.1). It can easily be shown that in a coordinate

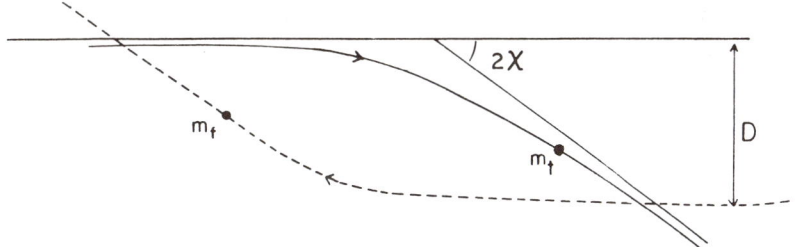

FIGURE AI.1. Geometry for encounter of two stars

system that moves with the center of gravity of test and field particle (center of mass system) the orbits of both particles are hyperbolas, the asymptotes of which make an angle $\pi - 2\chi$, with χ given by

(AI.1)
$$\sin \chi = \left[1 + \frac{D^2 V^4}{G^2 (m_t + m_f)^2} \right]^{-1/2}.$$

Thus the angular deflection of the test particle is 2χ in the center of mass system. In Equation (AI.1) D is the impact parameter, V is the relative velocity of the two particles, and m_t and m_f are the masses. To obtain the energy exchange in the encounter and the deflection in a coordinate system at rest with respect to the field particles we should transform the velocities before and after the encounter to that system. Since the velocity of the center of gravity is different in magnitude and direction in each new encounter this transformation is also different, and the evaluation of the average effect, though not difficult in principle, is quite cumbersome in practice. We shall therefore restrict ourselves to the case where the field particles are all at rest, with no random motions, and only state at the end of this appendix the modifications introduced by these random motions.

We first evaluate T_D by investigating the growth rate of the velocity component orthogonal to the velocity of the unperturbed test particle. The velocity of the center of gravity of the test and field particle is

$$(AI.2) \qquad v_g = \frac{m_t}{m_t + m_f} V,$$

and the velocity of the incoming test particle in the center of mass system thus is

$$(AI.3) \qquad v_t = \frac{m_f}{m_t + m_f} V.$$

Thus in an encounter we have for the increment in the perpendicular velocity component

$$(AI.4) \quad (\Delta v_\perp)^2 = \frac{m_f^2}{(m_t + m_f)^2} V^2 \sin^2(2\chi) = \frac{4 m_f^2}{(m_t + m_f)^2} V^2 \sin^2\chi \cos^2\chi.$$

If we consider an ensemble of test particles the mean value of $\Delta \mathbf{v}_\perp$ clearly vanishes because the direction of $\mathbf{v}_\perp$ in the plane orthogonal to $\mathbf{v}$ is random. But the mean of $(\Delta \mathbf{v}_\perp)^2$ does not vanish. If we consider N encounters the total $v_\perp^2$ acquired will be given by

$$(AI.5) \qquad v_\perp^2 = \left[\sum_{i=1}^{N} (\Delta \mathbf{v}_\perp)_i \right]^2 = \sum_{i=1}^{N} (\Delta v_\perp)_i^2 + \sum_{i,j=1; i \neq j}^{N} (\Delta \mathbf{v}_\perp)_i (\Delta \mathbf{v}_\perp)_j.$$

Taking the mean for the ensemble of test particles the products $(\Delta \mathbf{v}_\perp)_i (\Delta \mathbf{v}_\perp)_j$ all vanish if subsequent interactions are independent. Thus if we have N encounters per unit time the mean growth rate of $v_\perp^2$, which we denote by $\langle\langle v_\perp^2 \rangle\rangle$, is

(AI.6) $$\langle\langle \Delta v_\perp^2 \rangle\rangle = N(\Delta v_\perp)^2.$$

If we insert Equation (AI.1) in Equation (AI.4) we have

(AI.7)
$$(\Delta v_\perp)^2 = \frac{4 m_f^2 V^6 D^2}{G^2 (m_t + m_f)^4 [1 + D^2 V^4 / G^2 (m_t + m_f)^2]^2}$$
$$= \frac{4 m_f^2 V^2}{(m_t + m_f)^2} \frac{x^2}{(1 + x^2)^2}$$

with

$$x = \frac{D V^2}{G(m_t + m_f)}.$$

Since the result depends on the value of the impact parameter D we should multiply this expression with the number of encounters per unit of time with impact parameters between D and $D + dD$ and then integrate over all impact parameters to obtain $\langle\langle v_\perp^2 \rangle\rangle$. The number of encounters as a function of the impact parameter is

(AI.8) $$N(D)\,dD = 2\pi n\, VD\,dD = \frac{2\pi G^2 (m_t + m_f)^2 n}{V^3}\, x\,dx,$$

where n is the density of field stars. Thus we finally obtain

(AI.9) $$\langle\langle v_\perp^2 \rangle\rangle = \frac{8\pi G^2 m_f^2 n}{V} \int_0^\Lambda \frac{x^3}{(1 + x^2)^2}\, dx,$$

where Λ is the value of x that corresponds to the largest impact parameter that need be considered. For the integral we have

(AI.10) $$\int_0^\Lambda \frac{x^3\,dx}{(1 + x^2)^2} = \frac{1}{2} \ln(1 + \Lambda^2) - \frac{\Lambda^2/2}{1 + \Lambda} \approx \ln \Lambda \quad (\text{if } \Lambda \gg 1).$$

Thus if $\Lambda \to \infty$ the integral diverges. It might seem reasonable to choose a maximum impact parameter equal to the mean distance between stars, because for larger impact parameters the test particle interacts with many field particles at the same time, and it may be argued that in that case the deflections obtained from the theory of two-body encounters are too large. But Cohen, Spitzer and Routly have shown in a special case that the correct result is obtained by taking Λ much larger. They considered the situation in a plasma where the maximum impact parameter then should be taken equal to the Debye length (the length over which the effects of a particular charged particle are noticeable; at larger distances particles of opposite (charge) produce a shielding effect). Since gravitating particles all attract each other and thus no shielding occurs, this result would mean that we should integrate to the Λ that corresponds to an impact parameter equal to the size of the stellar system. The result is plausible for the following reason. From Equation (AI.7) it follows $(x \gg 1)$ that the effect of encounters at a distance D is proportional to m_f^2/D^2, and in the two-body approximation the "total effect" due to all particles within a radius D (the particles with which at a given instant an effective encounter takes place) is proportional to $(m_f^2/D^2) \, nD^3 = m_f^2 nD$. Clearly if D is large the two-body approximation is invalid. However, the total number of stars in the volume is nD^3, and thus statistical fluctuations of $n^{1/2}D^{3/2}$ in the number or $n^{1/2}D^{3/2}m_f$ in the mass may be expected. We can now consider the interaction of such a fluctuation with the test particle in the two-body approximation. Within the volume there is one such fluctuation at a mean distance of order D, but the effect is proportional to its mass squared divided by D^2, or $nD^3 m_f^2/D^2 = nm_f^2 D$. Thus the result obtained by considering these fluctuations is the same as that obtained by using the results of the two-body encounters well beyond their limit of validity. Thus the maximum impact parameter may be taken to be equal to the size of the system. With masses of a solar mass for test and field particles, $V = 10 \, \mathrm{km/sec}$ and $D_\mathrm{max} = 1000 \, \mathrm{pc}$ we would have $\ln \Lambda = 16$. If we had taken for D_max the mean distance between stars near the sun, say $3 \, \mathrm{pc}$, we would have had $\ln \Lambda = 10$, and thus the value of D_max does not matter too much. The quantitative results obtained thus are likely to be correct within a factor of two; nevertheless a more refined theory in which the transition from two-body encounters to the interactions with fluctuations is made in a satisfactory manner, is clearly desirable.

Returning to the determination of T_D it would seem that a mean deviation of the direction of motion of the test particle equal to $\pi/2$ corresponds about to $v_\perp^2 = V^2$. Thus we have for the time T_D

$$(AI.11) \qquad T_D = V^2 / \langle\!\langle v_\perp^2 \rangle\!\rangle = V^3 / (8\pi G^2 m_f^2 n \ln \Lambda).$$

Next we consider the energy exchange in one encounter. In the center of mass system the magnitude of the velocity of the test particle is unchanged. Thus in the center of mass system we have for the change in the velocity component parallel to the original velocity of the test particle

$$(AI.12) \quad \Delta v_\parallel = \frac{m_f}{m_t + m_f} V(\cos 2\chi - 1) = -\frac{2m_f}{m_t + m_f} V\sin^2\chi.$$

Since $\mathbf{v}$ and $\mathbf{v}_g$ are parallel this is also $\Delta v_\parallel$ in the rest frame. $(\Delta v_\perp)^2$ is again given by Equation (AI.7), and since the change in energy per unit mass ΔE is equal to $\frac{1}{2}(V + \Delta v_\parallel)^2 + \frac{1}{2}(\Delta v_\perp)^2 - \frac{1}{2}V^2$, we have

$$(AI.13) \quad \Delta E = -\frac{2m_t m_f}{(m_t + m_f)^2} V^2\sin^2\chi = -\frac{2m_t m_f}{(m_t + m_f)^2} \frac{V^2}{1 + x^2}.$$

With Equation (AI.8) for the number of encounters we have for the mean rate of change of the energy per unit mass of the test particle

$$(AI.14) \quad \langle\langle E \rangle\rangle = -\frac{4\pi G^2 m_t m_f n}{V} \int_0^\Lambda \frac{x\,dx}{1 + x^2} \approx -\frac{4\pi G^2 m_t m_f n}{V} \ln\Lambda.$$

Clearly the energy change is negative as the field particles had no kinetic energy. The time in which the change in E becomes about equal to $-E$ may be taken to be the time needed to establish equipartition of energy between a set of test particles and the field particles, if the masses are not too different. Thus we have

$$(AI.15) \quad T_{eq} = \frac{-E}{\langle\langle E \rangle\rangle} = \frac{V^3}{8\pi G^2 m_t m_f n \ln\Lambda} = \frac{m_f}{m_t} T_D.$$

If the mean kinetic energy of field particles and test particle had been equal, $\langle\langle E \rangle\rangle$ would vanish. In that case it would be more suitable to consider the change in E^2. Then we would have $\langle\langle E^2 \rangle\rangle = \overline{N}(\Delta E)^2$. Let us in general define a quantity $\langle\langle (\Delta E)^2 \rangle\rangle$ by the relation $\langle\langle (\Delta E)^2 \rangle\rangle = \overline{N}(\Delta E)^2$ where $\overline{N}$ is the number of encounters per unit time. If $\langle\langle E \rangle\rangle \neq 0$ this quantity is not equal to $\langle\langle E^2 \rangle\rangle$. We then find again with Equation (AI.8) for the number of encounters

$$(AI.16) \quad \langle\langle (\Delta E)^2 \rangle\rangle = \frac{8\pi G^2 m_t^2 m_f^2 n V}{(m_t + m_f)^2} \int_0^\Lambda \frac{x\,dx}{(1 + x^2)^2}.$$

Since the integral does not diverge (because the deviations in a complete encounter are first squared and only then averaged the distant encounters do not contribute much) we may take $\Lambda = \infty$. Then defining a time T_E we have

$$\text{(AI.17)} \qquad T_E = \frac{E^2}{\langle\!\langle (\Delta E)^2 \rangle\!\rangle} = \frac{(m_t + m_f)^2 V^3}{16\pi G^2 m_t^2 m_f^2 n} = \frac{(m_t + m_f)^2}{2m_t^2} \ln \Lambda \, T_D.$$

Finally we consider the dynamical friction. From Equations (AI.12) and (AI.8) we find

$$\text{(AI.18)} \qquad \langle\!\langle v_\parallel \rangle\!\rangle = -\frac{4\pi G^2 (m_t + m_f) m_f n}{V^2} \int_0^\Lambda \frac{x \, dx}{1 + x^2}.$$

Hence

$$\text{(AI.19)} \quad T_f = \frac{-V}{\langle\!\langle v_\parallel \rangle\!\rangle} = \frac{V^3}{4\pi G^2 (m_t + m_f) m_f n \ln \Lambda} = \frac{2m_f}{m_t + m_f} T_D.$$

Thus if we consider the cases $m_t \to 0$, $m_t = m_f$ and $m_f \to 0$ we have

$$m_t \to 0: \qquad T_E \to \frac{m_f^2}{2m_t^2} \ln \Lambda \, T_D; \quad T_{\text{eq}} \to \infty; \quad T_f \to 2T_D,$$

$$\text{(AI.20)} \; m_t = m_f: \qquad T_{\text{eq}} = T_f = T_D; \quad T_E = 2 \ln \Lambda \, T_D,$$

$$m_f \to 0: \qquad T_f \to 2T_{\text{eq}} = \frac{2m_f}{m_t} T_D \to \infty; \quad T_E \to \frac{1}{2} \ln \Lambda \, T_D \to \infty.$$

The general case where the field particles also move has been considered by Chandrasekhar and later by Spitzer. Chandrasekhar considers terms containing $\ln \Lambda$ dominant. This is usually the case, but in certain limiting cases the nondominant terms may well contribute appreciably since $\ln \Lambda$ is only of order 10. From the full results we can determine the correction factors that should be applied if the root mean square velocity of the field stars becomes large. The results are shown in Table AI.I. The ratio T_E / T_{E0} contains a factor $m_t^2 / (m_t + m_f)^2$, and in Table AI.I we

TABLE AI.I

The ratio of the times of relaxation for $\langle w^2 \rangle \neq 0$ to the case where $\langle w^2 \rangle = 0$ and the ratio T_D / T_E

$(\langle w^2 \rangle / V^2)^{1/2}$	T_D / T_{D0}	T_E / T_{E0}	T_f / T_{f0}	T_D / T_E
3	3.3	0.056	21	1.9
1	1.4	0.039	1.7	1.1
1/3	1.04	0.21	1.0	0.15

have taken $m_t = m_f$. Spitzer has considered also the important case in which both test particles and field particles have a Maxwellian velocity distribution, but with a different mean square velocity. The time scale on which equipartition is reached is obtained by averaging $\langle\langle \Delta E \rangle\rangle$ over both distributions. The result is

(AI.21) $$T_{eq} = (\langle w_t^2 \rangle + \langle w_f^2 \rangle) / (8(6\pi)^{1/2} G^2 m_t m_f n \ln \Lambda).$$

Results for T_D have already been tabulated in Table I. It should be noted here that V and n, which enter the time of relaxation, are not wholly independent in a self-gravitating stellar system. Let us consider a system with only random motions (E0 galaxy or globular cluster). The virial theorem states that in a mechanical system in a steady state the potential energy is equal to minus twice the kinetic energy. If we take a homogeneous sphere we thus have

(AI.22) $$(3/5)(GM^2/R) = M \langle V^2 \rangle.$$

Since $M \propto R^3 n$ we have

(AI.23) $$T_D \propto V^3/n \propto M^{1/2} R^{3/2} \propto n^{1/2} R^3.$$

Thus if in a given volume we introduce more and more mass, the increase in velocity dispersion is more important than the increase in density, and the relaxation time increases. A small radius is more important than a large density. At fixed density the time of relaxation goes down as R^3, and this is the reason of the low relaxation times in star clusters. Objections could be raised to the use of the steady state assumption in the application of the virial theorem to systems with long relaxation times. However, one should distinguish two kinds of relaxation in systems in which encounters are rare. Such systems generally will first attain a "rather" smooth and steady configuration on the time scale in which the stars can traverse the system a few times. Only much later, after a time T_D or T_E, will the systems attain complete relaxation also in velocity space.

It might be argued that because of the long mean free path (VT_D) in the gas of stars the viscosity should be large. However, for example, in the solar neighborhood the stars oscillate only with small amplitudes around their mean circular orbits around the galactic center. Thus a star never moves far in the radial direction, and the net momentum transfer if an encounter takes place is not large. Since encounters are rare these effects are negligible. The situation is analogous to that of a plasma in a magnetic field where the particles gyrate around a line of force. In general the concept of viscosity should only be used with great caution in a collision-free system.

APPENDIX II. The third-order moments of the Boltzmann equation

Of the third-order moment equations six contain nonvanishing terms. The others contain only moments odd in v_ϖ or v_z, which we have neglected. The equations are the following:

(AII.1)
$$\varpi \langle v_\varpi^2 \rangle \frac{\partial \langle v_\varpi^2 \rangle}{\partial \varpi} - 2u_\phi \langle v_\varpi^2 w_\phi \rangle + \langle v_\varpi^2 \rangle \langle w_\phi^2 \rangle$$
$$- \langle v_\varpi^2 w_\phi^2 \rangle + \frac{1}{n} \frac{\partial}{\partial \varpi} \left[\varpi n (\langle v_\varpi^4 \rangle - 3\langle v_\varpi^2 \rangle^2) \right] = 0,$$

(AII.2)
$$3 \langle v_z^2 \rangle n \frac{\partial \langle v_z^2 \rangle}{\partial z} + \frac{\partial}{\partial z} \left[n (\langle v_z^4 \rangle - 3\langle v_z^2 \rangle^2) \right] = 0,$$

(AII.3)
$$n \langle v_z^2 \rangle \frac{\partial \langle v_\varpi^2 \rangle}{\partial z} + \frac{\partial}{\partial z} \left[n (\langle v_\varpi^2 v_z^2 \rangle - \langle v_\varpi^2 \rangle \langle v_z^2 \rangle) \right] = 0,$$

(AII.4)
$$\frac{\langle v_\varpi^2 \rangle}{\varpi} \frac{\partial}{\partial \varpi} (\varpi^2 \langle w_\phi^2 \rangle) - 2\langle w_\phi^2 \rangle^2 - 2u_\phi \langle w_\phi^3 \rangle + 2\langle v_\varpi^2 w_\phi \rangle \frac{\partial (\varpi u_\phi)}{\partial \varpi} + 3\langle w_\phi^2 \rangle^2$$
$$- \langle w_\phi^4 \rangle + \frac{1}{n\varpi^2} \frac{\partial}{\partial \varpi} \left[\varpi^3 n (\langle v_\varpi^2 w_\phi^2 \rangle - \langle v_\varpi^2 w_\phi^2 \rangle) \right] = 0,$$

(AII.5)
$$\varpi \langle v_\varpi^2 \rangle \frac{\partial \langle v_z^2 \rangle}{\partial \varpi} - 2u_\phi \langle w_\phi v_z^2 \rangle - (\langle w_\phi^2 v_z^2 \rangle - \langle w_\phi^2 \rangle \langle v_z^2 \rangle)$$
$$+ \frac{1}{n} \frac{\partial}{\partial \varpi} \left[\varpi n (\langle v_\varpi^2 v_z^2 \rangle - \langle v_\varpi^2 \rangle \langle v_z^2 \rangle) \right] - 0,$$

(AII.6) $\langle v_z^2 \rangle \dfrac{\partial w_\phi^2}{\partial z} + 2\langle w_\phi v_z^2 \rangle \dfrac{\partial u_\phi}{\partial z} + \dfrac{1}{n} \dfrac{\partial}{\partial z} \left[n (\langle w_\phi^2 v_z^2 \rangle - \langle w_\phi^2 \rangle \langle v_z^2 \rangle) \right] = 0.$

With these equations we can derive in a simple way some results obtained in the literature from a solution of the Boltzmann equation under the assumption that stellar random motions obey a Schwarzschild distribution. We thus assume that the distribution function factorizes into three independent Maxwellian distribution functions. Then all odd moments would vanish and moreover

$$\langle v_\varpi^4 \rangle = 3\langle v_\varpi^2 \rangle^2$$

and

$$\langle v_\varpi^2 v_z^2 \rangle = \langle v_\varpi^2 \rangle \langle v_z^2 \rangle,$$

with similar relations for the other components. Equations (AII.1) and (AII.3) show that $\langle v_\varpi^2 \rangle$ is a constant, and Equations (AII.2) and (AII.5) show that $\langle v_z^2 \rangle$ is also constant. The last equation would give

$$\partial \langle w_\phi^2 \rangle / \partial z = 0.$$

Also Equation (53) would be exact and eliminating $\langle w_\phi^2 \rangle$ by this equation, Equation (AII.4) would become

(AII.7) $$\frac{1}{\varpi} \frac{\partial}{\partial \varpi} \left[\frac{\varpi^2}{2u_\phi} \frac{\partial (\varpi^2 u_\phi)}{\partial \varpi} \right] - \frac{\varpi}{2u_\phi^2} \left[\frac{\partial (\varpi u_\phi)}{\partial \varpi} \right]^2 = 0.$$

It can easily be seen that the solution of this equation is given by

(AII.8) $$u_\phi = C\varpi / (1 + D\varpi^2)$$

where C and D are integration constants. This rotation law was obtained by Chandrasekhar from the Boltzmann equation.

If, further, we take the z-derivative of Equation (46) and the ϖ-derivative of Equation (47), with $\partial u_\phi^2 / \partial z = 0$ from Equation (52) and $\partial \langle w_\phi^2 \rangle / \partial z = 0$ from the above discussion, we obtain $\langle v_\varpi^2 \rangle = \langle v_z^2 \rangle$. Both this conclusion and the rotation law (AII.8) are at variance with observation, and thus the assumption of a Schwarzschild-type distribution function leads to unacceptable results. Also Poisson's equation has not yet been used, and it can be shown that it is not compatible with the above results for a system of finite mass and density. Thus the Schwarzschild distribution should not be used in theoretical analysis except perhaps in some special cases.

Appendix III. Nearly circular orbits

When encounters between stars are unimportant, each star describes an independent orbit in the galactic gravitational field, and the study of these orbits is of interest. From a knowledge of the details of the orbits we can again infer properties of the distribution function of stellar velocities. Of course, the results that are obtained are also contained in the Boltzmann equation, but in many investigations complementary approaches prove useful.

In cylindrical coordinates the equations of motion for a star of unit mass in a gravitational field are ($\dot{\varpi} = d\varpi/dt$, etc.)

$$\ddot{\varpi} = \varpi\dot{\phi}^2 - \partial\Phi/\partial\varpi,$$

(AIII.1) $$\varpi\ddot{\phi} = -2\dot{\varpi}\dot{\phi} - (1/\varpi)\partial\Phi/\partial\phi,$$

$$\ddot{z} = -\partial\Phi/\partial z.$$

If Φ is axisymmetric the second equation yields upon integration $\varpi^2\dot{\phi} = J_z$, with J_z the constant angular momentum. Then we have

$$\ddot{\varpi} = J_z^2/\varpi^3 - \partial\Phi/\partial\varpi = -\partial\Psi/\partial\varpi,$$

(AIII.2) $$\ddot{z} = -\partial\Phi/\partial z = -\partial\Psi/\partial z,$$

with $\Psi = \Phi + \frac{1}{2}J_z^2/\varpi^2$, and thus the problem is two-dimensional.

Epicyclic orbits. We consider a nearly circular orbit by making use of a perturbation procedure, in which the zeroth order corresponds to a purely circular orbit in the galactic plane. Thus we have $\varpi_0 = \text{const}$, $\phi_0 = (\Theta/\varpi)(t - t_0)$ and $z_0 = 0$. In the nearly circular orbit we take $\varpi = \varpi_0 + \delta\varpi$, $\phi = \phi_0 + \delta\phi$ and $z = \delta z$. Then to the first order in the small quantities

$$\delta\ddot{\varpi} = \frac{J_z^2}{\varpi_0^3}\left(1 - 3\frac{\delta\varpi}{\varpi_0}\right) - \left(\frac{\partial\Phi}{\partial\varpi}\right)_0 - \left(\frac{\partial^2\Phi}{\partial\varpi^2}\right)_0\delta\varpi,$$

(AIII.3)

$$\delta\ddot{z} = -\left(\frac{\partial^2\Phi}{\partial z^2}\right)_0\delta z.$$

We choose ϖ_0 in such a way that the nearly circular orbit and the zeroth-order orbit have the same angular momentum; thus $J_z = \varpi_0\Theta$, and the first equation becomes

(AIII.4) $$\delta\ddot{\varpi} + \left[\frac{3}{\varpi_0}\left(\frac{\partial\Phi}{\partial\varpi}\right)_0 + \left(\frac{\partial^2\Phi}{\partial\varpi^2}\right)_0\right]\delta\varpi = 0.$$

Thus

$$\delta\varpi = a\sin\kappa(t - t_1),$$

$$\delta z = b\sin\beta(t - t_2),$$

with

(AIII.5) $$\kappa^2 = \frac{3}{\varpi_0}\left(\frac{\partial\Phi}{\partial\varpi}\right)_0 + \left(\frac{\partial^2\Phi}{\partial\varpi^2}\right)_0$$

and

$$\beta^2 = \partial \Phi / \partial z^2.$$

Here a and b are the arbitrary amplitudes; κ is the so-called epi-cyclic frequency. If $\kappa^2 < 0$ the orbit is clearly unstable, as $\delta \varpi$ grows exponentially. Thus stable circular orbits are only possible if $\partial \Phi / \partial \varpi$ decreases slower than ϖ^{-3}. If the force decreases faster than the inverse cube of ϖ no circular or nearly circular orbits can remain.

Since $J_z = \varpi_0 \Theta$, we have

$$\varpi_0 \Theta = \varpi^2 \dot{\phi} = \varpi_0^2 \dot{\phi}_0 + 2 \varpi_0 \dot{\phi}_0 \delta \varpi + \varpi_0^2 \delta \dot{\phi}$$

or

(AIII.6) $$\delta \dot{\phi} = - 2 (\Theta / \varpi_0^2) \, \delta \varpi,$$

or upon integration

(AIII.7) $$\delta \phi = (2 \Theta a / \varpi_0^2 \kappa) \cos \kappa (t - t_1).$$

In a local Cartesian coordinate system the origin of which rotates with the angular velocity Θ / ϖ_0 we thus have

$$x = a \sin \kappa (t - t_1),$$

(AIII.8) $$y = (2 a \Theta / \kappa \varpi_0) \cos \kappa (t - t_1),$$

$$z = b \sin \beta (t - t_2).$$

Thus the motions in the plane and transverse to it are uncoupled. In the plane the star describes an epicyclic orbit around a guiding center that moves with the local circular velocity. The epicyclic orbit is traversed in a direction opposite to that of the rotation, because the particle rotates most slowly when it is furthest from the center. The epicyclic orbit is elliptical with an axial ratio $2 \Theta / (\kappa \varpi_0)$. We have

(AIII.9) $$\kappa^2 = \frac{2 \Theta}{\varpi} \left(\frac{\Theta}{\varpi} + \frac{\partial \Theta}{\partial \varpi} \right) = 4 \Omega^2 + 2 \varpi \Omega \frac{\partial \Omega}{\partial \varpi} = 4 B (B - A),$$

with A and B the Oort constants of galactic rotation, and thus the axial ratio is equal to

$$2^{1/2} \left(1 + \frac{\varpi}{\Theta} \frac{\partial \Theta}{\partial \varpi} \right)^{-1/2} = \left(\frac{B - A}{B} \right)^{1/2}$$

or near the sun about 1.6. The period in the epicyclic orbit is

$$\text{(AIII.10)} \qquad T_{ep} = \frac{2\pi}{\kappa} = \frac{2\pi}{\left[\dfrac{2\Theta}{\varpi} \left(\dfrac{\Theta}{\varpi} + \dfrac{\partial \Theta}{\partial \varpi} \right) \right]^{1/2}} = \frac{\pi}{[B(B-A)]^{1/2}},$$

while the period of rotation around the galactic center is

$$\text{(AIII.11)} \qquad T_{circ} = \frac{2\pi}{\Theta/\varpi} = \frac{2\pi}{A-B}.$$

In a Newtonian force field $\Theta \propto \varpi^{-1/2}$ and the two periods are equal, corresponding to closed elliptical orbits in an inertial frame. Near the sun the epicyclical period is a factor 1.26 shorter than the rotation period, and thus the orbit is of the rosette type. Numerically we have near the sun

$$\begin{aligned}
T_{circ} &= 250 \times 10^8 y, \\
\text{(AIII.12)} \qquad T_{ep} &= 200 \times 10^8 y, \\
T_z &= 70 \times 10^8 y.
\end{aligned}$$

As a typical example of the usefulness of the epicyclic orbits, we discuss the evolution of expanding associations, that is, groups of very young stars that have apparently been ejected at a given time from a common point of origin. Let us thus consider a group of stars that at the time $t = 0$ were situated at the origin of a Cartesian coordinate system that rotates with the local circular velocity. At $t = 0$ all stars are given the same velocity V in random directions. Let θ be the angle between $\mathbf{V}$ and the positive x axis.

It is clear that though at $t = 0$ all stars were located at the same point, their guiding centers were not, and we have to take into account the difference in the angular velocity of the guiding centers and the origin of the coordinate system. If we consider a star whose guiding center is at $x = \eta$ we have, instead of Equation (AIII.8)

$$x = a \sin \kappa (t - t_1) + \eta,$$

$$\text{(AIII.13)} \qquad \begin{aligned}
y &= 2a\,(\Omega/\kappa) \cos \kappa (t - t_1) + \varpi_0 \left[\Omega(\eta) - \Omega(0) \right](t - t_2) \\
&= 2a\,(\Omega/\kappa) \cos \kappa (t - t_1) + \varpi_0 (\partial\Omega/\partial\varpi)\,\eta\,(t - t_2).
\end{aligned}$$

In this equation the four constants a, η, t_1 and t_2 should be determined from the initial conditions. In fact, at $t = 0$: $x = 0$, $y = 0$, $\dot{x} = V\cos\theta$, $\dot{y} = V\sin\theta$. Hence

$$\eta = a \sin \kappa t_1,$$

$$\varpi\,(\partial\Omega/\partial\varpi)\,\eta t_2 = (2a\Omega/\kappa)\cos\kappa t_1,$$

(AIII.14)
$$V\cos\theta = a\kappa\cos\kappa t_1,$$

$$V\sin\theta = 2a\Omega\sin\kappa t_1 + \varpi\,(\partial\Omega/\partial\varpi)\,\eta$$

$$= a(2\Omega + \varpi\,(\partial\Omega/\partial\varpi))\sin\kappa t_1 = (a\kappa^2/2\Omega)\sin\kappa t_1.$$

Expanding the sin and cos in Equation (AIII.14) and inserting the above results, we obtain

$$x = \frac{V}{\kappa}\left[\sin\kappa t\cos\theta + \frac{2\Omega}{\kappa}(1 - \cos\kappa t)\sin\theta\right],$$

(AIII.15)

$$y = \frac{2\Omega V}{\kappa^2}\left[-(1 - \cos\kappa t)\cos\theta + \left(\varpi\frac{\partial\Omega}{\partial\varpi}t + \frac{2\Omega}{\kappa}\sin\kappa t\right)\sin\theta\right].$$

Thus for any t the stars are situated on an ellipse whose major axis makes an angle χ with the x-direction, given by

(AIII.16)
$$\tan 2\chi = \frac{1 - \cos\kappa t}{\frac{1}{2}\varpi\frac{\partial\Omega}{\partial\varpi}t + \left(\frac{\Omega}{\kappa} + \frac{\kappa}{4\Omega}\right)\sin\kappa t}.$$

Thus the orientation of the ellipse gradually changes from the radial direction to a more tangential one, and from the orientation we can derive the age. The ages thus derived seem to be consistent with those from stellar evolution.

On the basis of the epicyclic theory the relationship between the rotational velocities of the stars and their random motions can also be discussed. Let us suppose that in the galactic plane the density of guiding centers of epicyclic orbits with radial amplitude a is an axisymmetric function $f(\varpi, a)$ and that the phases in the epicyclic orbits are random. Consider a ring with radius ϖ and width $d\varpi$. For the number of stars in this ring with guiding centers between $\varpi + \eta$ and $\varpi + \eta + d\eta$ we have, to the lowest order,

$$dN(\eta, a) \propto 2\pi\varpi f(\varpi, a)(1/\dot{x})\,d\varpi\,d\eta\,da.$$

From Equations (AIII.13) we have, to first order,

$$\dot{x}(\varpi) = \kappa(a^2 - \eta^2)^{1/2},$$

(AIII.17)

$$\dot{y}(\varpi) = \kappa^2\eta/2\Omega.$$

If we insert this in the expression for dN and obtain the normalization constant in dN from the condition that the total number of particles with given a per unit volume is $f(\varpi, a)\, da$, we have

(AIII.18) $\qquad dN(\eta, a) = 2\varpi f(\varpi, a)\, d\varpi\, d\eta\, da / (a^2 - \eta^2)^{1/2}$.

If we multiply a velocity component with $dN(\eta, a)$ and integrate over η from $-a$ to $+a$, we obtain the number of particles in the ring multiplied with the mean value of the velocity component, all for particles with a definite a. Integration over all a then results in the mean velocity over all particles. Thus to the first order in η we have

$$N\langle \dot{x}^2 \rangle = \kappa^2 \int_{-\infty}^{+\infty} da \int_{-a}^{+a} (a^2 - \eta^2)\, dN(\eta, a)\, d\eta = \pi \varpi \kappa^2 \int_{-\infty}^{+\infty} a^2 f(a)\, da,$$

(AIII.19)
$$N\langle \dot{y}^2 \rangle = \frac{\kappa^4}{4\Omega^2} \int_{-\infty}^{+\infty} da \int_{-a}^{+a} \eta^2 dN(\eta, a)\, d\eta$$
$$= \frac{\pi \kappa^4 \varpi}{4\Omega^2} \int_{-\infty}^{+\infty} a^2 f(a)\, da,$$

$$N\langle \dot{y} \rangle = \frac{\kappa^2}{2\Omega} \int_{-\infty}^{+\infty} da \int_{-a}^{+a} \eta\, dN(\eta, a)\, d\eta = 0.$$

Thus we obtain in first order

(AIII.20) $\qquad\qquad\qquad \theta - u_\phi - 0$

and

(AIII.21) $\qquad\qquad \dfrac{\langle \dot{y}^2 \rangle}{\langle \dot{x}^2 \rangle} = \dfrac{\langle w_\phi^2 \rangle}{\langle v_\varpi^2 \rangle} = \dfrac{\kappa^2}{4\Omega^2} = \dfrac{1}{2} \dfrac{\partial \ln(\varpi u_\phi)}{\partial \ln \varpi}$,

the latter in accordance with the result (53) derived from the Boltzmann equation. In the first order $\langle \dot{y} \rangle - 0$, but in the second order the relation (48) between $\dot{y}$ and $\langle v_\varpi^2 \rangle$ and $\langle w_\phi^2 \rangle$ is recovered.

The present type of analysis could be extended to situations with less symmetry. For example, if stars originate in spiral arms, the younger stars at least will have a nonaxisymmetric distribution of guiding centers, and the epicyclic phases may be not random. Such more elaborate theories may help us in understanding the vertex deviation, that is, the fact that the direction of largest velocity dispersion does not coincide with that of the galactic center for the younger stars.

Dispersion orbits. We consider the stellar orbit in a system of coordinates that rotates with arbitrary angular velocity Ω_c. We have in the rotating coordinates (ϖ', ϕ') instead of Equations (AIII.5) and (AIII.7)

(AIII.22)
$$\varpi' = \varpi_0' + a \sin \kappa (t - t_1),$$
$$\phi' = \phi_0' + (\Omega - \Omega_c)(t - t_1) + \frac{a}{\varpi_0'} \frac{2\Omega}{\kappa} \sin \kappa (t - t_1).$$

Thus to the first order in a/ϖ_0 the orbit in the rotating system is given by

(AIII.23)
$$\varpi' = \varpi_0' + a \sin \left[\frac{\kappa}{\Omega - \Omega_c} (\phi' - \phi_c') \right].$$

Let us choose Ω_c such that $\kappa/(\Omega - \Omega_c)$ is locally independent of ϖ_0'; thus

$$\frac{1}{\Omega - \Omega_c} \frac{\partial \kappa}{\partial \varpi} - \frac{\kappa}{(\Omega - \Omega_c)^2} \frac{\partial \Omega}{\partial \varpi} = 0$$

or

(AIII.24)
$$\frac{\kappa}{\Omega - \Omega_c} = \frac{d\kappa}{d\Omega}.$$

Then the orbit is called a dispersion orbit. For if we consider a cloud of matter scattered over a range of ϖ_0 (that is, with guiding centers scattered over this range) the cloud will disperse along this dispersion orbit by the galactic rotation, and two nearby pieces of matter disperse along orbits that are everywhere near. If $d\kappa/d\varpi$ is an integer, the dispersion orbit is closed and matter is "trapped" in the dispersion orbit. According to Lindblad, in most of the galaxy $\partial \kappa/\partial \Omega$ is either near 1 or near 2, and this could be of importance in the theory of spiral structure.

APPENDIX IV. Mass models of galaxies

We discuss the determination of the density distribution in a galaxy, from a knowledge of the rotation curve. We assume that the galaxy is axisymmetric and that the rotation velocities are equal to the circular velocities. Most rotation curves have been derived from measurements of Doppler shifts in emission lines in galaxies and thus refer to the gaseous component, and our assumptions may be reasonable unless magnetic fields or other perturbing effects are strong.

The basic assumption will be that the surfaces of equal density are oblate ellipsoidal surfaces of constant eccentricity. The light distribution in external galaxies indicates that this assumption is usually not unreasonable. The radial acceleration $f(\varpi,a)$, due to a homogeneous ellipsoid of unit density and semimajor axis a, at a point situated in the equatorial plane at a distance ϖ from the center of the ellipsoid (with eccentricity e) is, for $\varpi > a$, given by

$$f(\varpi,a) = 2\pi G\frac{(1-e^2)^{1/2}}{e^2}\left[\frac{a}{\varpi}\,(\varpi^2 \quad e^2a^2)^{1/2}\right.$$

(AIV.1)

$$\left. -\frac{\varpi}{e}\,\sin^{-1}\left(\frac{ea}{\varpi}\right)\right].$$

The radial acceleration due to a shell of matter between ellipsoidal surfaces with semimajor axes a and $a+da$ and density $\rho(a)$ is

(AIV.2) $$dF(\varpi,a) = \frac{\partial f(\varpi,a)}{\partial a}\,\rho(a)\,da,$$

and the radial acceleration due to a mass distribution with ellipsoidal equidensity surfaces of constant eccentricity $[\rho = \rho(a)]$ is

$$\frac{\Theta^2}{\varpi} = F(\varpi) = \int_0^\varpi \frac{\partial f(\varpi,a)}{\partial a}\,\rho(a)\,da$$

(AIV.3)

$$= \frac{4\pi G(1-e^2)^{1/2}}{\varpi}\int_0^\varpi \frac{\rho(a)a^2da}{(\varpi^2-e^2a^2)^{1/2}}.$$

The upper integration limit is ϖ since shells with $a > \varpi$ do not contribute (Newton's theorem). Thus if $\rho(a) = qa^n$ the substitution $a = (\varpi/e)\sin\theta$ yields

(AIV.4) $$\Theta^2 = 4\pi Gq\frac{(1-e^2)^{1/2}}{e^{n+3}}\varpi^{n+2}\int_0^{\sin^{-1}e}\sin^{n+2}\theta\,d\theta.$$

The total mass contained in the ellipsoid with $a = \varpi$ is in this case

$$M = 4\pi(1-e^2)^{1/2}\int_0^\varpi \rho(a)a^2da$$

(AIV.5)

$$= \frac{4\pi(1-e^2)^{1/2}q}{n+3}\varpi^{n+3}; \qquad (\text{for } n > -3).$$

Therefore

$$(\text{AIV.6}) \quad \Theta^2 = \frac{GM}{\varpi} \frac{n+3}{e^{n+3}} \int_0^{\sin^{-1}e} \sin^{n+2}\theta \, d\theta \qquad (n > -3).$$

If $e = 0$, the part that depends on n equals unity, and we recover the expression for a sphere. If $e = 1$ (flat disk) the integration limits are 0 and $\pi/2$, and the integral can be evaluated in terms of gamma functions. We obtain

$$(\text{AIV.7}) \quad \Theta^2(e = 1) = \frac{GM}{\varpi} \pi^{1/2} \frac{\Gamma((n+5)/2)}{\Gamma((n+4)/2)} \qquad (n > -3).$$

The ratio of Θ^2 for a disk to that for a sphere of the same mass is tabulated for some values of n in Table AIV.I.

TABLE AIV.I. Ratio of Θ^2 for a flat disk to Θ^2 for a sphere of the same mass

n	$\Theta^2(e = 1)/\Theta^2(e = 0)$
-2	$\pi/2$
-1	2
0	$3\pi/4$
$+1$	$8/3$
$+2$	$15\pi/8$
$\rightarrow \infty$	$(n\pi/2)^{1/2}$

A disk gives a larger attractive force than a sphere of the same mass because the matter is, on the average, nearer to the point where the force is evaluated. If the density increases inwards $(n < 0)$ the increase in the force for the disk compared to that for the sphere is less than 2.36, but if n is large and positive the difference increases. If $n \rightarrow \infty$ the mass distribution approaches that of a circular ring of zero thickness, which produces infinite force on its edge.

In our galaxy Θ is approximately constant between the sun and the region near the center. Thus if the mass distribution is represented by a single ellipsoid we have from Equation (AIV.4) that $n = -2$. Thus the density increases strongly inwards. It is satisfactory to note that this density gradient is intermediate between the density gradients of extreme stellar populations. From the run of the Θ's within the solar orbit, the mass situated in the spheroids with $a < \varpi$ can be reliably estimated. Outside the solar orbit Θ cannot be found from observation and only uncertain extrapolations have been made. Eventually an accurate evaluation of the velocity of escape near the sun might help us to estimate the total amount of matter outside the solar orbit. If one considers these uncertainties and the uncertainties in Θ_0 and ϖ_0, the total mass is rather uncertain. It probably is between one and two times 10^{11} solar masses.

In theoretical developments it is sometimes useful to consider the flat disk case because the z-coordinate can then be eliminated. Some care is needed to see of what configurations these disks represent the limiting case. Ellipsoids with $a > \varpi$ do not contribute to $F(\varpi)$. This remains true if $e = 1$. But the ellipsoids with $a > \varpi$ do contribute to the surface density at points inside the circle with radius ϖ. Thus if the surface density outside of this circle is changed F will be changed, if not at the same time the surface density within the circle is changed in a suitable manner. The relation between the surface density σ and $\rho(a)$ at a point a distance ϖ from the center is

(AIV.8)
$$\sigma(\varpi) = 2(1 - e^2)^{1/2} \int_\varpi^\infty \frac{\rho(a)\, a\, da}{(a^2 - \varpi^2)^{1/2}}.$$

A uniform disk with radius R is obtained for $e = 1$ if we take

$$\rho(a) = q(1 - e^2)^{-1/2}(R^2 - a^2)^{-1/2}.$$

Then $\sigma = \pi q$. Thus, according to Equation (AIV.3), we have for the radial force due to a uniform disk with radius $R(< \varpi)$

(AIV.9)
$$F(\varpi) = \frac{4G\sigma}{\varpi} \int_0^R \frac{a^2 da}{(\varpi^2 - a^2)^{1/2}(R^2 - a^2)^{1/2}}$$
$$= \frac{4G\sigma R^2}{\varpi^2} \int_0^{\pi/2} \frac{\sin^2\phi\, d\phi}{[1 - (R/\varpi)^2 \sin^2\phi]^{1/2}},$$

where $\sin\phi = a/R$. This is an elliptic integral that can be expressed in terms of the complete elliptic integrals of the first and second kind, K and E. We have

(AIV.10)
$$F(\varpi) = 4G\sigma(K\{R/\varpi\} - E\{R/\varpi\}).$$

Thus for a ring with radius R and width dR

(AIV.11)
$$dF = \frac{\partial F}{\partial R} dR = \frac{4G\sigma R dR}{\varpi^2 - R^2} E\left\{\frac{R}{\varpi}\right\}.$$

If $R > \varpi$ the upper limit of the integral in Equation (AIV.9) becomes ϖ instead of R; we take $\sin\phi = a/\varpi$ and find

(AIV.12)
$$F(\varpi) = 4G\sigma \frac{R}{\varpi}\left(K\left\{\frac{\varpi}{R}\right\} - E\left\{\frac{\varpi}{R}\right\}\right) \qquad (R > \varpi),$$

and for a ring with radius R and width dR

(AIV.13)
$$dF = -\frac{4G\sigma R dR}{\varpi^2}\left[\frac{E\{\varpi/R\}}{1 - \varpi^2/R^2} - K\left\{\frac{\varpi}{R}\right\}\right] \qquad (R > \varpi)$$

Let us approximate a galactic spiral arm as a circular ring. Near the sun the arms may be taken 500 pc wide and about 200 pc thick, with a density of 1H atom cm^{-3}. Thus $\sigma = 1.6 \times 10^{-3} g\, cm^{-3}$ (the mass per H atom is about $2.7 \times 10^{-27} g$ if helium and other elements are included) or about 15 percent of the total surface density. We consider a point 600 pc from the axis of the arm, where the disk approximation does not introduce too large errors. Making use of Equation (AIV.11) with $R/\varpi = 0.94$, we find $dF = 2 \times 10^{-10}$ c.g.s. compared with $\theta^2/\varpi = 2 \times 10^{-8}$ c.g.s. This changes the rotational velocity by 1.3 km/sec. A spiral arm is not axisymmetric. If the arm is inclined at an angle ψ to the azimuthal direction, a tangential acceleration $dF\sin\psi$ is present which would lead to a change in u_ϕ of 1.5 km/sec after a period of revolution of the galaxy, if $\psi = 0.1$ radian and if the location of the matter that is accelerated with respect to the spiral arm would not have changed. Thus if only the gas revealed by the 21-cm observations were present in the spiral arms, the direct gravitational effects would be small. However, there may be more gas, and also the density of low velocity stars may be rather significant in a spiral arm (compare the discussion by Dr. Lin).

Appendix V. Notations

In describing our galaxy we have made use of three coordinate systems (Figure 3):

1. A spherical polar coordinate system r, l, b, centered at the sun; r is the distance from the sun, l the galactic longitude counted

from the direction of the center in a sense opposite to the galactic rotation, and b is the galactic latitude, which is the angle a given direction makes with the galactic plane. The north galactic pole is at $b = +90°$.

Galactic coordinates defined this way are frequently written as l^{II}, b^{II} to distinguish them from old galactic coordinates (in use before 1960), l^I, b^I, which were counted from a different zero point (the center at $l^I = 327°$), and in which the galactic plane did not accurately coincide with the plane $b^I = 0$.

2. A right-handed cylindrical coordinate system ϖ, ϕ, z, centered at the galactic center; ϖ is the distance from the center, ϕ the polar angle measured in the same direction as l and with the zero point chosen in such a way that the directions $l = 0$ and $\phi = 0$ coincide, and z is the height above the galactic plane, counted positive towards the north galactic pole. It is convenient to refer to ϕ as the galactocentric longitude.

3. A right-handed Cartesian coordinate system x, y, z, centered at the sun with the galactic center in the negative x-direction, the galactic rotation in the negative y-direction and the north galactic pole in the positive z-direction.

Velocity components have been denoted as v with a subscript referring to the particular coordinate direction. Similarly, the mean systematic velocity and the random velocity have been denoted as u and w with appropriate subscripts. It is not uncustomary in the stellar dynamical literature to write Π, Θ or Φ, Z for w_ϖ, w_ϕ, w_z, or R, Φ, Θ for w_r, w_ϕ, w_θ. Other symbols we have made use of include the following:

$\left. \begin{matrix} A \\ B \end{matrix} \right\}$ Oort constants of galactic rotation (Equation 62)

D — impact parameter

E — energy integral

$E_0 - E_7$ — elliptical galaxies

G — gravitational constant

H — magnetic field strength or Hamiltonian

I — integral of motion

J — angular momentum integral

P_{ij} — pressure tensor

S0	disk-like galaxies without spiral structure
Sa, Sb, Sc	spiral galaxies
SB0, SBa, SBb, SBc	barred spiral galaxies
a	amplitude of epicyclic motion or major axis of ellipsoid
f	distribution function of stellar velocities
m	mass of a star
n	number density of stars
p, q	coordinates in phase space
Θ	circular velocity
Φ	gravitational potential (force $= -\nabla\Phi$)
Ψ	density of representative points in phase space
β	angular frequency of oscillation in z-direction
κ	epicyclic frequency
ρ	mass density
σ	surface mass density

References

1. A. Blaauw and M. Schmidt (Editors), *Stars and stellar systems.* Vol. 5: *Galactic structure,* Univ. of Chicago Press, Chicago, Ill., 1965.

2. *Handbuch der Physik.* Vol. 53: *Stellar systems* edited by S. Flügge, Springer, New York, 1960.

3. S. Chandrasekhar, *Stellar dynamics,* Univ. of Chicago Press, Chicago, Ill., and Dover, New York, 1942.

4. L. Spitzer, *Physics fully ionized gases,* Interscience, New York, 1962.

5. L. Woltjer (Editor), *Interstellar matter in galaxies,* Benjamin, New York, 1963.

6. A. Sandage (Editor), *Hubble atlas of galaxies,* Carnegie Institute of Washington, 1962.

7. D. J. K. O'Connell (Editor), *Stellar populations,* Pontifical Academy of Sciences, Rome, 1958.

8. Detailed references to recent literature can be found in *Galactic structure* [1].

COLUMBIA UNIVERSITY
NEW YORK, NEW YORK

C. C. *Lin*

Stellar Dynamical Theory
of Normal Spirals

1. **Introduction.** In the concluding section of Woltjer's article in these Lectures, he discussed the two possible ways of identifying a spiral arm: (a) a material arm associated with a tube of gas, or (b) a wave pattern. He also explained the difficulties associated with the first possibility.[1] Basically, this is the difficulty of the material arm being wound up rather rapidly by the differential rotation of the system. The various attempts to avoid this difficulty lead to other uncomfortable assumptions: such as an essential dependence on the role of the halo, and the existence of rather large magnetic fields (5×10^{-5} gauss).

I shall describe in this lecture and the next, a theory of spiral structure based on possibility (b): i.e., the structure is basically associated with a wave pattern. As we shall see, this does not remove the difficulty of differential rotation immediately, for young stars are also the markers of a spiral arm. I shall therefore begin my discussion with a description of the observational aspects of the problem and then try to show that a *density wave*, if presumed to exist, would be adequate to account for the main observational

---

[1] For a detailed discussion of this problem see Oort [12].

features. This will be followed by a brief description of the theory for its formation and a report on the basic nature and the characteristics of this density wave according to our theory. Finally, I shall describe (mostly by the use of Appendices) some of the details of the analysis which were used to arrive at these conclusions.

For the sake of clarity, I shall describe only a line of approach adopted by myself.[2] There are other related investigations during the past few years, which I regret I shall not be able to discuss in detail. These will be briefly referred to from time to time. (See also the list of references to such work given at the end of this paper.)

2. **Statement of the problem.** Let us now review briefly some of the observational facts to see what the problems are.

1. The spiral arms are associated with the gas and the young stars. The contrast of mass density between the brighter and darker regions is small.

2. There is a typical spacing for spiral galaxies of various types as determined by other physical characteristics as shown in the following table:

$$Sa \rightarrow Sb \rightarrow Sc$$

Nuclear concentration	**decreasing**
Gas content	increasing
Arm spacing	increasing
Total mass	decreasing

3. There appears to be a considerable degree of regularity of the pattern *extending over the whole disk*. In the case of some of the galaxies, such as NGC 5364 or the whirlpool galaxy, this regularity is remarkably strong.

4. There is also the extraordinary "3-kpc arm" which we should bear in mind, although it might not be possible to account for it in a first attempt to construct a theory.

We are therefore faced with the following two problems:

(a) Why do most of the disk-like galaxies have the regular structure?

(b) How can these structures persist in the presence of differential rotation?

---

[2] Mr. Frank H. Shu collaborated in this work.

The problem of persistence is very well stated by Oort [12] in the following manner:

"In systems with strong differential rotation, such as is found in all nonbarred spirals, spiral features are quite natural. Every structural irregularity is likely to be drawn out into a part of a spiral. But *this* is not the phenomenon we must consider. We must consider a spiral structure extending over the whole galaxy, from the nucleus to its outermost part, and consisting of two arms starting from diametrically opposite points. Although this structure is often hopelessly irregular and broken up, the general form of the large-scale phenomenon can be recognized in many nebulae."

In other words, the primary problem is to explain the *grand design* over the whole disk on the scale of 10 kpc. There is also a secondary problem of describing the structure of an individual *arm*, which might involve somewhat different mechanisms.

In looking for mechanisms for creation and maintenance of regular spiral patterns, it should be borne in mind that one could also use essentially the same mechanism to account for spiral patterns which are *less perfect* and comparatively transitory. For the superposition of several patterns rotating relative to each other would have the desired effect. On the other hand, if we have only a theory for producing spiral arms, one would still have to explain by what mechanism are these spiral arms organized into a regular pattern.

3. **Kinematical considerations.** Since the first difficulty that we face is a kinematical one, let us begin by examining the consistency of our concept with the observed kinematical behavior of the various components of our galaxy. The concept of a density wave pattern must pass two "kinematical tests": the predictions on (i) the radial motion of the gas and (ii) the winding of young stellar arms, must both be consistent with observations.

The behavior of the older stars in a galaxy can be readily observed only in terms of a statistical distribution and therefore does not present any problem in a physical picture involving a density wave. A similar remark applies to the gas, but here we have already a more serious point to be reconciled. The *material* velocity of the gas can be observed by the Doppler effect, and it is necessary to show that the radial motion of the gas required by a density wave, circulating around the galactic center, does not

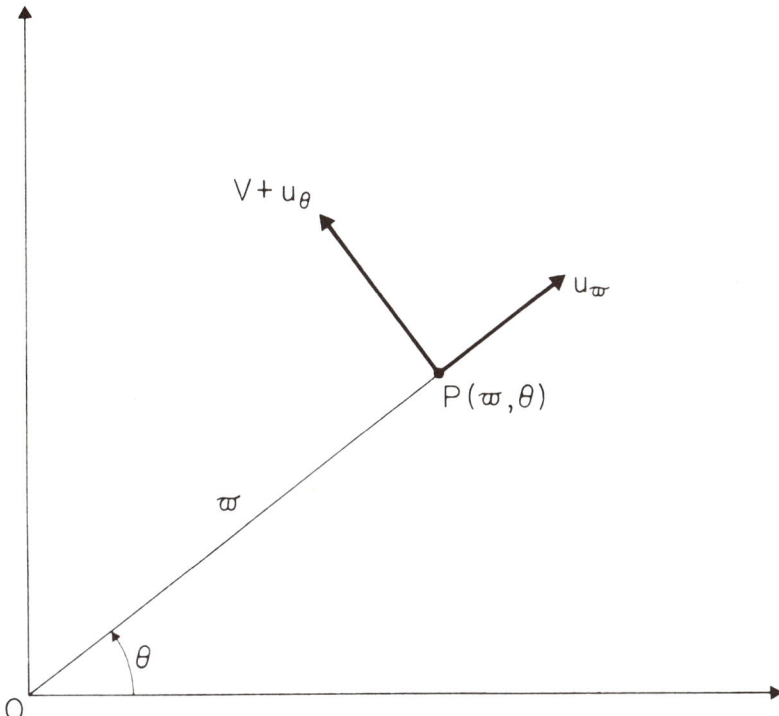

FIGURE 1. Coordinate system in the galactic disk

exceed any upper limit that might be imposed by observations, and we shall see that this is indeed the case in our own galaxy.

The arrangement of young stars can also be followed in greater detail. It is possible, at least approximately, to draw "isochronic lines" for young stars of the same age. If the winding process of differential rotation is operative, then these arms of young stars should be rapidly wound up in the course of one or two revolutions. However, the "life time" of these brilliant young stars of the O and B types are of the order of ten million years, and we shall see that the result of the winding process over such a short period is not impressive.

Consider, therefore, an observer moving with the angular velocity Ω_p of the spiral pattern. The density pattern of the gas will then be stationary, but the gas particles will be moving at a different speed. Let us denote the gas velocity by $(u_\varpi, V + u_\theta)$, (see Figure 1); then it can be shown, by considering the equation of continuity, that the radial motion of the gas has an upper limit $u_{\max}$ given by

$$(3.1) \qquad \frac{u_{\max}}{V(\varpi)} = \frac{1}{\pi} \frac{\Delta\varpi}{\varpi} \frac{1-r}{1+r}\left(1 - \frac{\Omega_p}{\Omega}\right)$$

where $\Delta\varpi$ is the spacing between two successive arms around the location ϖ, r is the ratio of minimum surface density of the gas to its maximum surface density, and Ω is the angular velocity at ϖ. If we take

$$r = 1/3, \qquad \Delta\varpi = 2\,\text{kpc at } \varpi = 5\,\text{kpc,}$$

$$\Omega = 50\,\text{km/sec-kpc,} \qquad \Omega_p = 20\,\text{km/sec-kpc,}$$

$$V = 250\,\text{km/sec,}$$

we obtain

$$(3.2) \qquad\qquad u_{\max} = 8\,\text{km/sec.}$$

This is safely within the upper limit of 15 km/sec set by observations (see [8, p. 8]).

Consider now the behavior of the young stars. They are presumed to be born in the gas concentration ACB (Figure 2). After

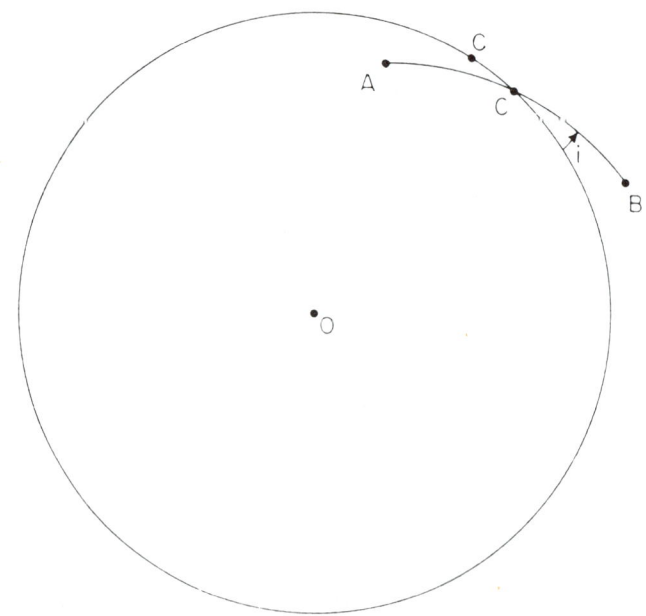

FIGURE 2. Motion of young stars relative to the gaseous arm

a time interval Δt, the stars born at C would move *ahead* of the density wave (which is still at ACB) to the point C' given by

$$\overset{\frown}{C'C} = \varpi\,(\Omega - \Omega_p)\,\Delta t.$$

The stars born at A would advance further while the stars born at B would not advance as much so that the arc $A'C'B'$ would be inclined at a somewhat smaller angle i' ($< i$). But the stars in the arc $A'C'B'$ do not deviate very much from the arc ACB. Indeed, the deviation of C' from ACB is given by

(3.3) $$\delta\varpi = (\Omega - \Omega_p)\,(\Delta t)\tan i.$$

If we take the above values of ϖ, Ω and Ω_p, and $\Delta t = 10 \times 10^6$ yrs., $\tan i = 0.11$, we get

(3.4) $$\delta\varpi = 170\,\mathrm{pc}.$$

Thus, the stellar arm is now *slightly outside* of the gaseous arm. There is indeed observational evidence to give general support to this kind of displacement ([20], [15] and Becker, 1963). Detailed comparative study of observational data and theoretical predictions would be desirable.

The change of the angle of inclination is clearly very small over ten million or even twenty million years. For longer periods of time, the stars merge into the general population and can no longer be distinguished as a spiral arm. In this way, the winding dilemma is avoided as far as the young stellar arms are concerned.

4. **Dynamical processes.** Having passed the "kinematical tests," we now turn to the dynamical basis of these density waves. They represent "cooperative" behavior of the stars due to gravitational interaction. The existence of these density waves was suggested by the late B. Lindblad many years ago, and he had tried continuously, for a period of about 25 years, to develop a theory for it. Unfortunately, his method depends so much on the study of individual stellar orbits[3] that it is extremely difficult to compare his theory with our own, —which is based on a statistical treatment through the use of the distribution function.

Before we proceed with a purely gravitational theory, we should

---

[3] P. O. Lindblad obtained density waves by large-scale machine calculation by using about 200 stars to simulate the galaxy. The number appears to be too small to give conclusive answers, although the results are indicative of the general mechanism.

consider all the components of the galaxy: stars, gas, magnetic field, photons, cosmic ray particles, etc., and make an assessment of their relative importance in a dynamical process. This may be done by using the basic equations of stellar dynamics and hydromagnetics. I shall not go into the details here. Suffice it to say (see Appendix I) that the hydromagnetic effect becomes appreciable for a scale of 1 kpc only if the field is as high as 5×10^{-5} gauss (which is uncomfortably high). Since we are dealing with the problem of the *spiral pattern*, on the scale of 10 kpc, it will never be as large as the gravitational forces involved.[4]

We shall therefore consider a purely gravitational theory, including both the gravitational field of the stars and of the gas. We attempt to demonstrate that density waves with a spiral structure will be self-sustained at a small but finite amplitude. Suppose such a wave were maintained, then there must be an associated gravitational field of a generally spiral form. We shall start with this field and carry out the analysis as indicated in the following diagram:

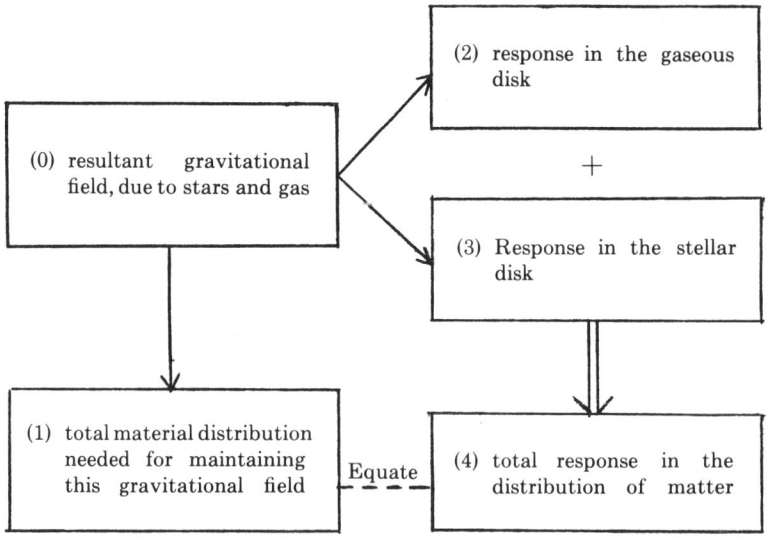

The resultant gravitational field (0) must be associated, according to Poisson's equation, with a certain distribution of matter (1),

---

[4] In this connection it might be mentioned that attempts to construct a spiral pattern theory through the use of hydromagnetic waves have not been successful, according to investigations made by D. Wentzel (unpublished).

which may consist partly of gas and partly of stars. The distribution of gas (2) may be calculated in terms of the *resultant* gravitational field without any *further* reference to the distribution of the stars. Similarly, the distribution of the stars (3) may be calculated without any *further* reference to the distribution of the gas. The sum of these two distributions yields a total distribution of matter (4) that must be identical with the density distribution (1) which is needed to give rise to the field. This last condition is the equation to be solved for the unspecified functions and parameters that occur in the resultant gravitational field (0) initially assumed.

The analysis has so far been carried out only for the linear theory. Although the nonlinear effects may be expected to tend to prefer trailing patterns, such a preference will be seen to occur already in the linear theory.

5. **The asymptotic theory: neutral waves.** To carry out the analysis, we adopt the natural cylindrical coordinate system (ϖ, θ, z) such that the galactic disk is in the plane $z = 0$, with its center at the origin. In the linear theory, the gravitational potential (item (0) above) may be assumed to be given by a superposition of spiral modes, and the response to these individual modes may be treated separately. Let the potential of each of these modes be given by

$$(5.1) \qquad \mathcal{V}_1(\varpi, \theta, t) = A(\varpi) \exp\{i[\omega t - m\theta + \Phi(\varpi)]\},$$

where $A(\varpi)$ and $\Phi(\varpi)$ are real functions of the axial distance ϖ, ω is a complex parameter, and m is a positive integer. The function $A(\varpi)$ is slowly varying with ϖ, whereas the function $\Phi(\varpi)$ is of the form $\epsilon^{-1}\phi(\varpi)$, where $\phi(\varpi)$ is slowly varying, and ϵ is a small parameter of the order of the angle of inclination i of the spiral arms. Indeed, the function (5.1) clearly has a spiral structure described by the family of curves

$$(5.2) \qquad m(\theta - \theta_0) = \Phi(\varpi) - \Phi(\varpi_0)$$

which has an angle of inclination i given by

$$(5.3) \qquad [\varpi \Phi'(\varpi)]^{-1} = (\tan i)/m.$$

Thus, a natural approach is to adopt an asymptotic solution based on a rapidly varying phase angle for all the functions of the general form (5.1).

(1) *The density distribution according to Poisson's equation.* It is easy to show that the surface density distribution associated

with the potential (5.1) has the form

$$(5.4) \qquad \sigma_1(\varpi,\theta,t) = s(\varpi,\epsilon)\exp\{i[\omega t - m\theta + \Phi(\varpi)]\},$$

where

$$(5.5) \qquad s(\varpi,\epsilon) = -|\Phi'(\varpi)|\{A(\varpi)/2\pi G\}\{1 + O(\epsilon)\}.$$

The explicit form of the higher order terms will be given elsewhere.

(2) *The density response in the gaseous disk*. From the gas-dynamical equations, a simple calculation shows that, in the linear approximation, the density distribution induced in the gaseous disk by the gravitational potential (5.1) is of the form

$$(5.6) \qquad \sigma_{\mathrm{ind}} = s_{\mathrm{ind}}(\varpi,\epsilon)\exp\{i[\omega t - m\theta + \Phi(\varpi)]\},$$

where

$$(5.7) \qquad s_{\mathrm{ind}} = \sigma_0 \frac{-|\Phi'(\varpi)|^2 A(\varpi)}{\kappa^2 - (\omega - m\Omega)^2}\{1 + O(\epsilon)\}.$$

In the above formula, $\sigma_0(\varpi)$ is the basic density distribution of the gas, $\varpi\Omega(\varpi)$ is the circular velocity, and $\kappa(\varpi)$ is the epicyclic frequency. This type of calculation was essentially used by Lin and Shu [5] in their elementary theory of self-sustained density waves over a galactic disk.

(3) *The density response in the stellar disk*. The calculation of the response of the stellar disk can be made by using the equations of stellar dynamics, but the detailed steps are naturally more complicated. These calculations will be published elsewhere; the basic equations and an outline of the essential steps may be found in Appendix II. Suffice it to say that, to the lowest approximation in the small parameter ϵ described above, and with the peculiar velocity of the stars satisfying the Schwarzschild distribution, a formula of the form (5.7) still holds, provided that the density distribution σ_0 is now replaced by an *equivalent stellar density* σ_e which is reduced from the true stellar density $\sigma_*(\varpi)$ by a suitable *reduction factor which is real for real* ω. That is, we must replace σ_0 in (5.7) by

$$(5.8) \qquad \sigma_e = \sigma_* \mathscr{F}_\nu(x)$$

where

$$(5.9) \qquad \mathscr{F}_\nu(x) = \frac{1-\nu^2}{x}\left\{1 - \frac{\nu\pi}{\sin\nu\pi}\mathscr{G}_\nu(x)\right\},$$

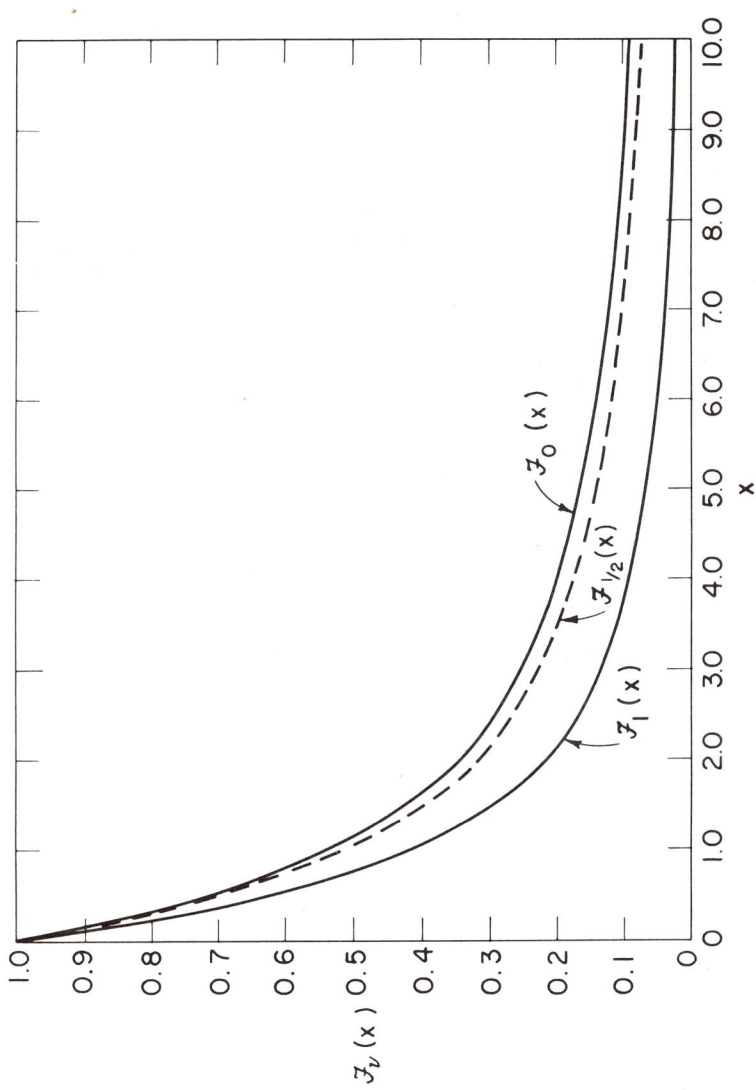

FIGURE 3. The reduction factor

and

(5.10) $\mathscr{G}_\nu(x) = \dfrac{1}{2\pi} \displaystyle\int_{-\pi}^{\pi} \cos(\nu s)\exp[-x(1 + \cos s)]ds.$

In the above formulae,

(5.11) $\nu = (\omega - m\Omega)/\kappa,$

(5.12) $x = |\Phi'(\varpi)|^2 \langle c_\varpi^2 \rangle / \kappa^2,$

and $\langle c_\varpi^2 \rangle$ is the mean square peculiar radial velocity of the stars in the basic symmetric distribution.

For real values of ν, $\mathscr{F}_\nu(x)$ is real, and its numerical values are shown in Figure 3, for $\nu = 0$, $|\nu| = 1$, and $|\nu| = \frac{1}{2}$ (dashed). For intermediate values of ν, a linear interpolation in ν^2 is a good approximation. A more complete set of numerical values is given in Table I. It is through the use of such a reduction factor that Lin and Shu [5, Remark 3, p. 652] discussed the effect of velocity dispersion in their elementary theory.

If we now follow the procedure in §4, we find the following formula for specifying the spiral pattern (5.2):

(5.13) $|\Phi'(\varpi)| = [\kappa^2 - (\omega - m\Omega)^2]/2\pi G[\sigma_0 + \sigma_* \mathscr{F}_\nu(x)].$

We note that the density distribution of the gas can only play a secondary role, at least at such locations typified by the interior part of our galaxy, where σ_0/σ_* may be as small as 2%. Even in our neighborhood where the total density is comparatively small, the gas can only have an influence comparable to that of the low speed stars.[5]

For a given distribution of the mean quantities such as $\Omega(\varpi)$, $\sigma_0(\varpi)$ and $\sigma_*(\varpi)$, and a given distribution of the velocity dispersion of the stars, the formula (5.13) gives a definite wave number $|\Phi'(\varpi)|$ for the radial spacing, provided that the angular velocity ω/m is given. Conversely, by taking the observed inter-arm spacing, one can estimate the velocity dispersion of the stars at each location.

(4) *Application to a galactic model.* Suppose we are given a basic state. Then the parameters Ω, κ, and $\langle c_\varpi^2 \rangle$ are known. Thus, according to Equation (5.13), to each real value of ω there is a determination of $|\Phi'(\varpi)| = |k(\varpi)|$ which gives the form of the spiral pattern for a range of values of ϖ such that

---

[5] In such cases, one should apply a similar "reduction factor" to the gaseous density, although the theory is less well founded. However, since the turbulent velocity of the gas is small, this is not expected to be a significant correction.

TABLE I. Numerical values of $\mathscr{F}_\nu(x)$

ν \ x	0	1	2	3	4	5	6	7	8
0	1	0.5342	0.3457	0.2523	0.1982	0.1633	0.1389	0.1209	0.1068
0.1	1	0.5333	0.3447	0.2514	0.1974	0.1625	0.1381	0.1202	0.1066
0.2	1	0.5305	0.3416	0.2484	0.1946	0.1600	0.1359	0.1182	0.1046
0.3	1	0.5258	0.3363	0.2434	0.1901	0.1558	0.1321	0.1147	0.1013
0.4	1	0.5191	0.3286	0.2362	0.1835	0.1499	0.1267	0.1097	0.0966
0.5	1	0.5100	0.3185	0.2266	0.1748	0.1420	0.1195	0.1032	0.0907
0.6	1	0.4984	0.3056	0.2145	0.1638	0.1321	0.1105	0.0950	0.0833
0.7	1	0.4838	0.2894	0.1994	0.1503	0.1199	0.0995	0.0849	0.0739
0.8	1	0.4658	0.2695	0.1810	0.1337	0.1050	0.0861	0.0726	0.0629
0.9	1	0.4435	0.2452	0.1586	0.1136	0.0871	0.0700	0.0581	0.0494
1.0	1	0.4158	0.2153	0.1312	0.0894	0.0656	0.0507	0.0408	0.0341

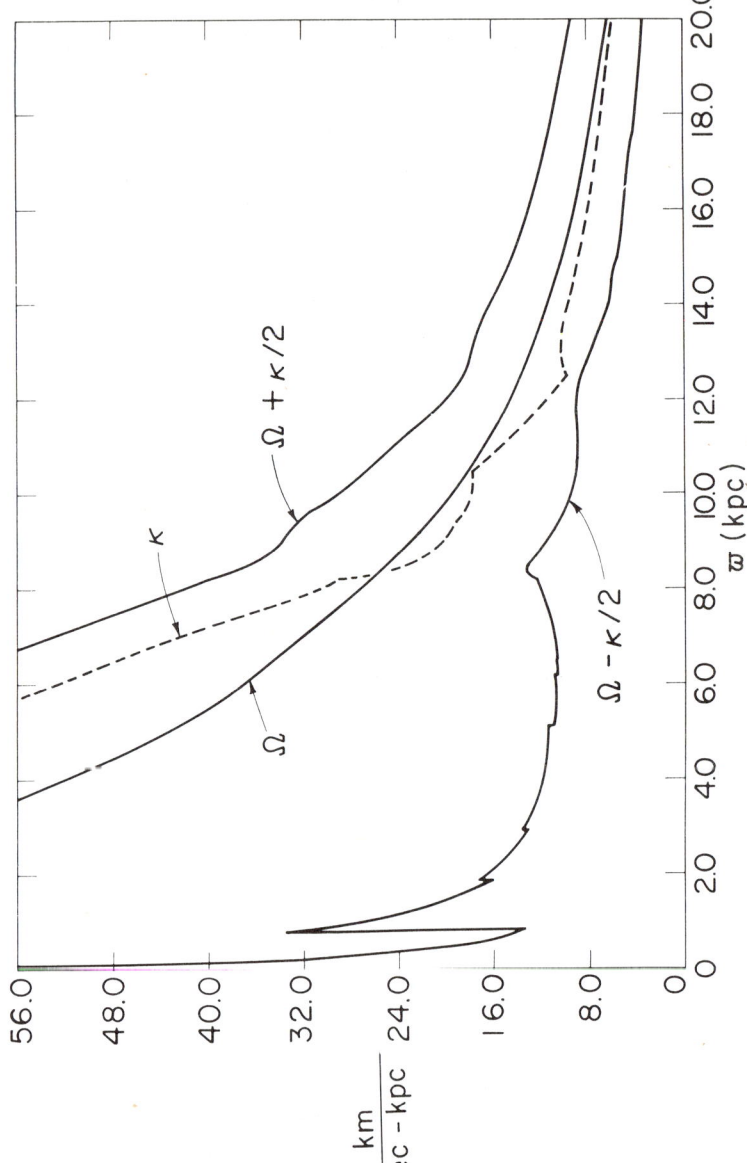

FIGURE 4. Angular velocity etc. of the Schmidt [16] model of our galaxy

(5.14) $$\kappa^2 - (\omega - m\Omega)^2 > 0.$$

If we now use the Schmidt model (for which the sun is located at 8.2 kpc), and take

(5.15) $$\Omega_p = \omega/2 = 20 \, \text{km/sec kpc},$$

then we find that the range defined by (5.14) extends over the radial distance (cf. Figure 4)

(5.16) $$2 \, \text{kpc} < \varpi < 12 \, \text{kpc}.$$

The inner radius is rather uncertain, and could easily be at 3 kpc. We shall refer to this range defined by (5.14) as the *principal part of the spiral pattern*.

Since we do not have any reliable data on $\langle c_\varpi^2 \rangle$, we could assume a value of the wave-number k for a given location ϖ, and calculate, according to (5.13), the amount of "participating mass," $\sigma + \sigma_* \mathscr{F}_\nu(x)$, and thereby determine the velocity dispersion $\langle c_\varpi^2 \rangle$. It turns out, by assuming a radial spacing between arms of about 2 kpc, that the fraction of stellar mass participating is of the order of 10—15% (in the Schmidt model). This means that the gravitational effect of the gas is definitely important in our neighborhood, but is not as significant toward the interior. At $\varpi = 4 \, \text{kpc}$, the gas contributes to perhaps 2% of the mass whereas the remaining "participating mass" of 8% must be attributed to the stars. From this, it is estimated that

(5.17) $$x = k^2 \langle c_\varpi^2 \rangle / \kappa^2 = 4.5.$$

This places the velocity dispersion in the "stable part" of Toomre's [18] diagram. It should be emphasized that this part of his diagram represents neutral stability permitting oscillations. The statement holds, of course, only under the approximations assumed. It can be easily verified that the initial approximation used in our asymptotic procedure is essentially equivalent to his "local approximation" although there are differences in emphasis and interpretation. Indeed, by putting $\nu = 0$, one can easily obtain a "stability criterion" comparable to Toomre's, but it depends on the ratio σ_0/σ_* (which was taken to be zero in his calculations).

We note that, up to the present approximation, *there is no way to differentiate between leading waves and trailing waves*. This is indeed expected from general symmetry considerations. In the next approximation, however, we shall find that these waves are indeed *overstable*. The instability is found to be dependent on certain quantities which are naturally neglected in a local approximation, such as

$$\frac{d\log\Omega}{d\log\varpi}, \quad \frac{d\log\langle c_\varpi^2\rangle}{d\log\varpi} \quad \text{and} \quad \frac{d\log\sigma_*}{d\log\varpi}.$$

6. **Amplication of trailing waves.** All the results above are stated for the initial approximation. In the next approximation, in which we include terms of the order of the angle of inclination of the spiral arms, we shall find indications of *instability*. This small amplification rate is then expected to be balanced[6] by the nonlinear effects to produce neutral waves at small but finite amplitudes. The amplitude at equilibrium is expected to be of the order of $\epsilon^{1/2}$, where $\epsilon = (1/m)\tan i$.

An indication of instability may already be found in the asymmetrical deviation from the Schwarzschild distribution, even if the small spiral disturbance[7] from the equilibrium distribution is calculated only to the initial approximation. It is found that this asymmetry leads to an amplification of trailing waves which may be described largely in terms of the "lag velocity" of the mean circular motion of the stars. By using the moment equations of stellar dynamics, it can be easily shown that the lag velocity is given by

$$(6.1) \quad -\langle c_\theta\rangle = \frac{\langle c_\varpi^2\rangle}{2\varpi\Omega}\left\{-\frac{1}{2}\frac{d\log\Omega}{d\log\varpi} + \frac{d\log\sigma_*}{d\log\varpi} + \frac{d\log\langle c_\varpi^2\rangle}{d\log\varpi}\right\}.$$

Thus, the effect of this term is indeed of the order of all the other effects of the nonlocal character. Since all the three quantities $\Omega(\varpi)$, $\sigma_*(\varpi)$ and $\langle c_\varpi^2\rangle$ are decreasing functions of ϖ, and $\langle c_\theta\rangle$ is negative, we see that the *amplification of trailing waves* is encouraged by the gradients of σ_* and $\langle c_\varpi^2\rangle$, whereas differential rotation per se has the *opposite* effect.

If we carry out the calculations indicated by Equation (41), Appendix II, we find further confirmation of the above conclusions. The numerical details have not been completely worked out, but asymptotic solutions for large x have been obtained. Besides the

---

[6] See Landau and Lifschitz, *Fluid dynamics*, p. 104 for a corresponding discussion of hydrodynamic stability.

[7] Cf. Equations (24), (33), and (41), Appendix II.

terms in $\langle c_\theta \rangle$, the calculation of $\phi^{(1)}$ yields an amplification rate $-\omega_i$ given by

$$(6.2) \qquad \frac{\omega_i}{\kappa} \, \mathrm{sign}(k) = \frac{1}{\kappa\varpi} \left\{ \frac{2\langle c_\varpi^2 \rangle}{\pi} \right\}^{1/2} f(\xi) \cdot D,$$

to the lowest approximation, where

$$(6.3) \qquad f(\xi) = \frac{\xi \sin \xi}{\sin \xi \cos \xi - \xi} > 0 \quad (\text{for } -\pi < \xi < 0),$$

$$(6.4) \qquad \xi = \frac{\omega_r - m\Omega}{\kappa} \, \pi,$$

and

$$(6.5) \qquad D = -\frac{d \log \langle c_\varpi^2 \rangle}{d \log \varpi} > 0.$$

This formula applies at a location ϖ where $A'(\varpi) = 0$. At other values of ϖ, $A(\varpi)$ varies slowly in such a manner that ω_i can maintain the same value.

The amplification index is given by $-\omega_i$, and trailing waves correspond to $k < 0$; the quantity appearing in (6.2) should be positive for the desired result. This is true for the range of ξ indicated in (6.3). In our own galaxy, this indicates the range

$$(6.6) \qquad\qquad 3 \, \mathrm{kpc} < \varpi < 10 \, \mathrm{kpc}$$

in the Schmidt model. Although the calculation shows that there might be a preference of leading waves in the range 10—12 kpc, this cannot be taken on face value. It should be borne in mind that this is already well beyond the solar neighborhood ($\varpi = 8.2$ kpc), and that this less massive part is surely to be driven essentially by the gravitational field and the motion of the more massive interior part. A slow variation of $A(\varpi)$ would then preserve the amplification of trailing waves in this locality.

The effect of differential rotation per se does not appear in the approximation (6.2) but in a term of a smaller order. Its effect is again to prefer leading waves against trailing waves, at least for those travelling near the circular speed of the stars.

Density waves travelling essentially with the material speed have been considered by Lynden-Bell and Goldreich [10] for a gaseous sheet and by Julian and Toomre [3] for the stellar sheet. In both investigations, differential rotation is included, while the gradients of σ_*

and $\langle c_{\varpi}^2 \rangle$ are not. It was found that leading wave patterns would be amplified and sheared out into trailing patterns, which are eventually smoothed out by phase mixing, in the case of the stellar sheet. This is not incompatible with our findings, although it is not immediately obvious whether the results are directly comparable.

7. **Concluding remarks.** Observations of the phenomena over a galactic disk show transient features as well as features which appear to be quasi-permanent. It is not difficult to account for transient features with generally spiral-like appearance[8] in a system with differential rotation. The central problem is indeed to account for the more *permanent* features such as the spiral pattern over the whole disk with characteristic spacing between spiral arms. In this paper, we have described such a theory in terms of the concept of density waves, and demonstrated the basic mechanism for its sustained existence in terms of gravitational forces. Comparison with observations shows that the numerical values thus obtained are within the expected accuracy (of the order of 20%, loosely speaking).

The theory has several important implications. One conclusion most directly related to observations is the prediction of gas motion varying periodically as

$$\genfrac{}{}{0pt}{}{\cos}{\sin} 2(\theta - \Omega_p t).$$

In the framework of our theory, the 3 kpc arm is related to a strong Lindblad resonance at this location, where

$$\Omega_p = \Omega - \kappa/2.$$

The gas is not moving *outwards* at all angular directions. Rather it is moving *inwards* in the two quadrants which have so far escaped detailed observations. In general, the theory suggests that one should reexamine the manner in which the observed 21 cm data are used to locate the distance of the gaseous arms. Since this determination is based on the Doppler shift observed and the assumption is usually made that the gas is in *exactly circular* motion, the gaseous

---

[8] For example, it might be suggested that the branching of spiral arms, which includes arms inclined at rather large angles with the circular direction, is indeed a transient feature, and that it might be related to the transient wave patterns considered by Lynden-Bell and Goldreich and by Julian and Toomre.

motion predicted here might lead to a significant change in the final assignment of the location of the gaseous arm. These and other astronomical implications will be examined in subsequent investigations.

APPENDIX I

DYNAMICAL EQUATIONS FOR A GALAXY OF STARS

For convenience of reference we collect below all the basic dynamical equations governing the behavior of all the three components of a galaxy: (i) the stars, (ii) the interstellar gas, and (iii) the magnetic field.

(A) *Equations of stellar dynamics.* Consider an idealized model with stars all having the same mass m_*. We introduce a distribution function Ψ_3 in the phase space (x_i, v_i), $(i = 1, 2, 3)$. The equation governing the development of Ψ_3 in time t is as follows:

$$(1) \qquad \partial \Psi_3 / \partial t + v_i \, \partial \Psi_3 / \partial x_i + a_i \, \partial \Psi_3 / \partial v_i = 0,$$

where v_i is the velocity of the individual stars, and a_i is the acceleration due to the combined gravitational field of the gas and the stars. The latter is thus derivable from a gravitational potential $\mathscr{V}(x_i, t)$:

$$(2) \qquad a_i = - \partial \mathscr{V} / \partial x_i,$$

and the gravitational potential is in turn determined by the gas and stars in the galaxy (and perhaps the effect of neighboring galaxies):

$$(3) \qquad \partial^2 \mathscr{V} / \partial x_k^2 = 4\pi G \rho_t, \qquad \rho_t = \rho + \rho_*.$$

In this equation, G is the gravitational constant, ρ is the mass density of the gas, and $\rho_*(x_i, t)$ is the mass density of the stars given by

$$(4) \qquad \rho_*(x_i, t) = m_* \int \Psi_3(x_i, v_i, t) \, d\tau(v_i),$$

where the integral extends over the infinite velocity space. The mass density of the gas is one of the variables in the hydromagnetic equations.

(B) *Equations of hydromagnetics.* Because of the large linear dimensions involved, the effect of viscosity and resistivity are expected to be small. With this provision, the hydromagnetic equations are as follows.[9] For the dynamics of the gas we have respectively the following equations of state, mass conservation, motion, and energy:

(5) $$p = \rho RT,$$

(6) $$\partial \rho / \partial t + \partial (\rho u_k) / \partial x_k = 0,$$

(7) $$\rho \frac{Du_i}{Dt} = -\frac{\partial p}{\partial x_i} + \rho a_i + \frac{1}{4\pi} \frac{\partial}{\partial x_j} \left\{ B_i B_j - \frac{1}{2} B^2 \delta_{ij} \right\},$$

(8) $$\frac{Dp}{Dt} = \gamma \frac{p}{\rho} \frac{D\rho}{Dt} + (\gamma - 1) \dot{q}.$$

The symbols in the above equations are as follows. The thermodynamic state of the gas is specified by the pressure p, the density ρ, and the temperature T. These are related through the parameter R, the gas constant per unit mass. The motion of the gas is described by the Eulerian velocity field $u_i(\mathbf{x}, t)$ and its substantial derivative Du_i/Dt, which is the acceleration. The force terms on the right-hand side of (7) contain not only the pressure force, but also (i) *the acceleration due to the combined gravitational field of the gas and the stars, and* (ii) *the force due to the Maxwell stresses of the magnetic field B_i*. These last two items are the agents through which the gas interacts with the other two important components of the galaxy: the stars and the magnetic field. The last equation for energy contains the unspecified term $\dot{q}$ which represents the rate of energy transfer into a unit volume per second by heat conduction, radiative transfer, etc.

The displacement current is usually neglected in a hydromagnetic problem. In the case of infinite conductivity, it is well known that the Maxwell equations lead to the following two equations for the magnetic field:

(9) $$\partial B_k / \partial x_k = 0,$$

(10) $$\frac{D}{Dt} \left(\frac{B_i}{\rho} \right) = \frac{B_k}{\rho} \frac{\partial u_i}{\partial x_k}.$$

The second equation above guarantees the conservation of magnetic

---

[9] Gaussian (c.g.s.) units are used throughout this paper.

flux associated with a given piece of fluid surface. (The magnetic field is "frozen" into the fluid.)

In order to get an idea of the relative importance of the various types of energy or force in the galaxies, we list below some typical values for our own galaxy:

TABLE II. Energy densities in the Galaxy[10]

Total radiation (star light)	$0.7 \times 10^{-12} \, \mathrm{erg \, cm^{-3}}$
Turbulent gas motion	$0.5 \times 10^{-12} \, \mathrm{erg \, cm^{-3}}$
Total energy of galactic rotation	$1300 \times 10^{-12} \, \mathrm{erg \, cm^{-3}}$
Cosmic rays	$1 \times 10^{-12} \, \mathrm{erg \, cm^{-3}}$
Magnetic field (10^{-5} gauss)	$4 \times 10^{-12} \, \mathrm{erg \, cm^{-3}}$

Clearly, the kinetic energy of galactic rotation far outweighs any other component. Furthermore, this table is based on a mean gas density of about 2×10^{-24} g/cc (1.2 H atoms/cc). The total mass density of gas and stars is about 5 times larger in the solar neighborhood, and about 50 times larger in the interior of the galaxy. It is thus clear that even an extremely small deviation of the stars from circular motion would involve a level of energy comparable to the rest. A one-percent deviation would be associated with substantially more energy density than, e.g., the total magnetic energy density.

While we wish to emphasize this predominance of stellar mass in our galaxy, the situation may not be as favorable in other galaxies, such as those of the Sc type which has considerably more gas. We have therefore formulated the basic equations to include all the three components named above. The effect of star light and cosmic rays will not be explicitly considered, even though they are acknowledged as the source for ionizing the gas.

We should note that the magnetic field might become more important if we consider *forces* instead of energy densities. Thus, if the field should vary substantially over the scale of one kpc, it would be ten times as important as the kinetic energy of galactic rotation at a location 10 kpc from the galactic center. Thus, a magnetic field of 5×10^{-5} gauss would yield an "equivalent" energy density of $1000 \times 10^{-12} \, \mathrm{erg \, cm^{-3}}$ in the above table.

---

[10] P. Morrison, Rev. Mod. Phys. **29** (1957), 235.

Appendix II

Mathematical Theory of Stellar Dynamics for an Infinitesimally Thin Disk of Stars in Differential Rotation

Notation.

t — time

ϖ, θ, z — cylindrical coordinates

Π, Θ, Z — components of total stellar velocity in cylindrical coordinates

$\Psi_3(\varpi, \theta, z, \Pi, \Theta, Z, t)$ — distribution function in "three-dimensional" case

$\Pi_0(\varpi, \theta, z, t), \qquad \Theta_0(\varpi, \theta, z, t), \qquad Z_0(\varpi, \theta, z, t)$

components of velocity for an "arbitrarily chosen" reference system ($\Pi_0 = Z_0 = 0$, $\Theta_0 = \varpi \Omega(\varpi)$, in the present paper)

$c_\varpi = \Pi - \Pi_0, \qquad c_\theta = \Theta - \Theta_0, \qquad c_z = Z - Z_0$

components of "peculiar velocity" relative to the reference system

In the "two-dimensional" case, the following notation is adopted:

$\Psi_2(\varpi, \theta, \Pi, \Theta, t)$ — distribution function in "two-dimensional case"

$$\sigma_* = m_* \int \int \Psi_2 d\Pi d\Theta$$

surface density, where m_* is the mass of an individual star

$\Pi = \Pi_0(\varpi, \theta, t) + c_\varpi, \qquad \Theta = \Theta_0(\varpi, \theta, t) + c_\theta$

with definition of symbols similar to those above

$$\langle q \rangle = \int \int q \Psi_2 d\Pi d\Theta \Big/ \int \int \Psi_2 d\Pi d\Theta,$$

local mean value for q.

$$\langle \Pi \rangle = \Pi_0 + \langle c_\varpi \rangle = \Pi_0 + v_\varpi = V_\varpi + v_\varpi,$$

$$\langle \Theta \rangle = \Theta_0 + \langle c_\theta \rangle = \Theta_0 + v_\theta = V_\theta + v_\theta,$$

$$[V_\varpi \text{ usually taken to be zero}, V_\theta = \varpi \Omega(\varpi)]$$

$\mathscr{V}(\varpi, \theta, z, t)$ — gravitational potential

a_ϖ, a_θ — acceleration components in the plane

$$\Psi(\varpi, \theta, c_\varpi, c_\theta, t) = \Psi_2(\varpi, \theta, \Pi, \Theta, t)$$

representation of the "two-dimensional" distribution function in terms of the peculiar velocities for a given choice of the reference system.

$$c'_\varpi = c_\varpi - \langle c_\varpi \rangle, \qquad c'_\theta = c_\theta - \langle c_\theta \rangle$$

components of peculiar velocity relative to the instantaneous mean motion

1. **Basic equations.** The basic equation of stellar dynamics, when written in terms of cylindrical coordinates (ϖ, θ, z), reads as follows [1, p. 187]:

$$
\begin{aligned}
(1) \quad & \frac{\partial \Psi_3}{\partial t} + \Pi \frac{\partial \Psi_3}{\partial \varpi} + \frac{\Theta}{\varpi} \frac{\partial \Psi_3}{\partial \theta} + Z \frac{\partial \Psi_3}{\partial z} \\
& - \left(\frac{\partial \mathscr{V}}{\partial \varpi} - \frac{\Theta^2}{\varpi} \right) \frac{\partial \Psi_3}{\partial \Pi} - \left(\frac{1}{\varpi} \frac{\partial \mathscr{V}}{\partial \theta} + \frac{\Pi \Theta}{\varpi} \right) \frac{\partial \Psi_3}{\partial \Theta} - \frac{\partial \mathscr{V}}{\partial z} \frac{\partial \Psi_3}{\partial Z} = 0,
\end{aligned}
$$

where t is the time, (Π, Θ, Z) are the components of the stellar velocity, $\mathscr{V}(\varpi, \theta, z, t)$ is the gravitational potential, and Ψ_3 is the distribution function such that

$$(2) \qquad dN = \Psi_3(\varpi, \theta, z, \Pi, \Theta, Z, t)\, \varpi\, d\varpi\, d\theta\, dz\, d\Pi\, d\Theta\, dZ$$

gives the number of stars (all of mass m_*) in the element of phase space

$$(3) \qquad d\tau = \varpi\, d\varpi\, d\theta\, dz\, d\Pi\, d\Theta\, dZ.$$

Let us now specialize to the problem of a "flat" disk. We introduce the *two-dimensional* distribution function

$$(4) \qquad \Psi_2(\varpi, \theta, \Pi, \Theta, t) = \int_{-\infty}^{\infty} \int_{-\infty}^{\infty} \Psi_3\, dz\, dZ,$$

and apply the same integration process to (1). Clearly, the terms involving Z disappear if Ψ_3 is such that (4) is absolutely convergent. A little calculation then shows that, in the limit of zero thickness, we have the following equation for an *infinitesimally thin disk*:

$$
\begin{aligned}
(5) \quad & \frac{\partial \Psi_2}{\partial t} + \Pi \frac{\partial \Psi_2}{\partial \varpi} + \frac{\Theta}{\varpi} \frac{\partial \Psi_2}{\partial \theta} \\
& + \left(a_\varpi + \frac{\Theta^2}{\varpi} \right) \frac{\partial \Psi_2}{\partial \Pi} + \left(a_\theta - \frac{\Pi \Theta}{\varpi} \right) \frac{\partial \Psi_2}{\partial \Theta} = 0.
\end{aligned}
$$

The surface density of this disk is

(6)
$$\sigma_*(\varpi,\theta,t) = m_* \int\!\!\int_{-\infty}^{\infty} \Psi_2 \, d\Pi \, d\Theta.$$

Due to a given surface density, such as (6), the components of acceleration (a_ϖ, a_θ) are given by

(7)
$$a_\varpi = -\left(\frac{\partial \mathcal{V}}{\partial \varpi}\right)_{z=0}, \qquad a_\theta = -\frac{1}{\varpi}\left(\frac{\partial \mathcal{V}}{\partial \theta}\right)_{z=0},$$

where $\mathcal{V}$ satisfies the Poisson equation

(8)
$$\frac{\partial^2 \mathcal{V}}{\partial \varpi^2} + \frac{1}{\varpi}\frac{\partial \mathcal{V}}{\partial \varpi} + \frac{1}{\varpi^2}\frac{\partial^2 \mathcal{V}}{\partial \theta^2} + \frac{\partial^2 \mathcal{V}}{\partial z^2} = 4\pi G \sigma_* \delta(z).$$

In (8), G is the gravitational constant, and $\delta(z)$ is Dirac's delta function.

We are usually interested in a system in differential rotation. Thus, it is convenient to introduce the following *peculiar velocity relative to an "arbitrarily" chosen reference state* $(\Pi_0 = 0, \Theta_0 = \varpi\Omega(\varpi))$:

(9)
$$c_\varpi = \Pi - 0, \qquad c_\theta = \Theta - \varpi\Omega(\varpi).$$

The components of mean "peculiar velocities," $\langle c_\varpi \rangle$ and $\langle c_\theta \rangle$, are *not* necessarily zero;[11] but it is implied that they are "small" compared with $\varpi\Omega(\varpi)$. We shall now introduce (c_ϖ, c_θ) as the independent variables in place of (Π, Θ) in (5), and write

(10)
$$\Psi(\varpi,\theta,c_\varpi,c_\theta,t) = \Psi_2(\varpi,\theta,\Pi,\Theta,t).$$

Then, Equation (5) becomes

(11)
$$\begin{aligned}
&\frac{\partial \Psi}{\partial t} + c_\varpi \frac{\partial \Psi}{\partial \varpi} + \left(\Omega + \frac{c_\theta}{\varpi}\right)\frac{\partial \Psi}{\partial \theta} \\
&\quad + \left(a_\varpi + \Omega^2\varpi + 2\Omega c_\theta + \frac{1}{\varpi}c_\theta^2\right)\frac{\partial \Psi}{\partial c_\varpi} \\
&\quad + \left(a_\theta - \frac{\kappa^2}{2\Omega}c_\varpi - \frac{1}{\varpi}c_\varpi c_\theta\right)\frac{\partial \Psi}{\partial c_\theta} = 0
\end{aligned}$$

where Ψ is associated with the volume element

---

[11] Usually, $\langle c_\varpi \rangle = 0$ and $\langle c_\theta \rangle < 0$, in the basic symmetrical state of a disk galaxy.

$$dT_c = \varpi \, d\varpi \, d\theta \, dc_\varpi \, dc_\theta$$

of the phase space, and

(12) $$\kappa^2 = (2\Omega)^2 \{ 1 + (\varpi/2\Omega)(d\Omega/d\varpi) \}.$$

The (local) averages of the components of the peculiar velocity are in general different from zero, and these shall be denoted by v_ϖ and v_θ, respectively. We have

(13)
$$\sigma_* v_\varpi = m_* \int \int c_\varpi \Psi \, dc_\varpi \, dc_\theta,$$

$$\sigma_* v_\theta = m_* \int \int c_\theta \Psi \, dc_\varpi \, dc_\theta,$$

where σ_* is the surface density:

(14) $$\sigma_* = m_* \int \int \Psi \, dc_\varpi \, dc_\theta$$

and m_* is mass per star. The local mean of the *total* stellar velocity (Π, Θ) is easily verified to be given by $(\varpi \Omega(\varpi) + v_\varpi, v_\theta)$ as expected.

Equation (11) is the final *equation for stellar dynamics with reference to a conveniently chosen differentially rotating system, for the configuration of a thin disk with infinitesimal thickness.* The acceleration (a_ϖ, a_θ) derives part of its contributions from the surface density σ_* given by (14), (according to (7) and (8)), but it may also have contributions from any "external" agency (e.g., gas or another neighboring galaxy).

2. **Equations for small disturbances of a certain symmetrical distribution.** Consider first a certain state of axially symmetric distribution *in equilibrium*, which we shall describe in terms of a distribution function

(15) $$\Psi = \Psi_0(\varpi, c_\varpi, c_\theta).$$

The components of acceleration are

(16) $$(a_\varpi, a_\theta) = (a_0(\varpi), 0).$$

The basic Equation (11) becomes

(17)
$$c_\varpi \, \partial \Psi_0/\partial \varpi + \{ a_{\varpi 0} + \Omega^2 \varpi + 2\Omega c_\theta + c_\theta^2/\varpi \} \partial \Psi_0/\partial c_\varpi$$
$$+ \{ 0 - (\kappa^2/2\Omega + c_\theta/\varpi) c_\varpi \} \partial \Psi_0/\partial c_\theta = 0.$$

For convenience, we shall choose $\Omega(\varpi)$ *such that the condition*

(18) $a_{\varpi\,0} + \Omega^2 \varpi = 0$

is satisfied. With this choice, the Equation (17) is simplified into

(19) $c_\varpi \dfrac{\partial \Psi_0}{\partial \varpi} + \left(2\Omega + \dfrac{c_\theta}{\varpi}\right) c_\theta \dfrac{\partial \Psi_0}{\partial c_\varpi} - \left(\dfrac{\kappa^2}{2\Omega} + \dfrac{c_\theta}{\varpi}\right) c_\varpi \dfrac{\partial \Psi_0}{\partial c_\theta} = 0.$

The general solution of this equation is of the form

(20) $\Psi_0 = \mathscr{F}_0\{\varpi\,(\varpi\Omega + c_\theta), \tfrac{1}{2}\,[c_\varpi^2 + (\varpi\Omega + c_\theta)^2] + \mathscr{V}_0(\varpi)\,\},$

which is a general function of the two integrals of angular momentum and energy:

(21) $J = \varpi\,(\varpi\Omega + c_\theta),$

(22) $E = \tfrac{1}{2}\,[c_\varpi^2 + (\varpi\Omega + c_\theta)^2] + \mathscr{V}_0(\varpi).$

In the above equations,

(23) $\mathscr{V}_0(\varpi) = \displaystyle\int_{\varpi_0}^{\varpi} \Omega^2 \varpi\, d\varpi.$

Note that $\mathscr{V}_0(\varpi)$ need not be due to the stellar distribution (15) alone, as there might be, for example, a symmetrical distribution of gas in the galaxy.

Consider now a slight deviation from Ψ_0 associated with a small change in (a_ϖ, a_θ). The distribution function is now

(24) $\Psi = \Psi_0(1 + \psi),$

and the field [which is *not* necessarily wholly determined by (24)] is

(25) $a_\varpi = a_{\varpi\,0} + a_{\varpi\,1}, \qquad a_\theta = a_{\theta 1}.$

We are thus considering the *change in* Ψ *due to an imposed field* $(a_{\varpi\,1}, a_{\theta 1})$ *while leaving it open whether the field is due to self-gravitation alone.* To be sure, there is a self-field $(a_\varpi^{(s)}, a_\theta^{(s)})$ associated with (24), and this may be subtracted from (a_ϖ, a_θ) to determine the contributions of other agencies (e.g., gas and other nearby galaxies).

To derive the differential equation for Ψ, it is convenient to write the fundamental equation (11) in the form

(26) $\mathscr{L}\,(\Psi) = 0$

where $\mathscr{L}$ is the linear differential operator defined by

$$\mathscr{L} \equiv \frac{\partial}{\partial t} + c_\varpi \frac{\partial}{\partial \varpi} + \left(\Omega + \frac{c_\theta}{\varpi}\right) \frac{\partial}{\partial \theta}$$

(27)
$$+ \left\{ a_\varpi + \Omega^2 \varpi + 2\Omega c_\theta + \frac{c_\theta^2}{\varpi} \right\} \frac{\partial}{\partial c_\varpi}$$

$$+ \left\{ a_\theta - \left(\frac{\kappa^2}{2\Omega} + \frac{c_\theta}{\varpi}\right) c_\varpi \right\} \frac{\partial}{\partial c_\theta}.$$

We shall denote the corresponding operator with $(a_\varpi, a_\theta) = (a_{\varpi 0}, 0)$ by $\mathscr{L}_0$ and write

(28)
$$\mathscr{L} = \mathscr{L}_0 + \mathscr{L}_1$$

where

(29)
$$\mathscr{L}_1 = a_{\varpi 1} \frac{\partial}{\partial c_\varpi} + a_{\theta 1} \frac{\partial}{\partial c_\theta}$$

and $(a_{\varpi 1}, a_{\theta 1})$ defined by (25) are "small" in some sense (to be clarified below). If we now substitute (24) into (26), we obtain, with the help of these new symbols and the relation $\mathscr{L}_0\{\Psi_0\} = 0$, the following *nonlinear equation for the small disturbance* ψ:

(30)
$$\mathscr{L}_0\{\psi\} = \mathscr{L}_1\{Q_0\} + [\psi\mathscr{L}_1\{Q_0\} - \mathscr{L}_1\{\psi\}]$$

where

(31)
$$Q_0 = -\log \Psi_0.$$

Since ψ is expected to be of the same order as the perturbation $(a_{\varpi 1}, a_{\theta 1})$ that causes it, the last two terms in (30) are of smaller order than the term $\mathscr{L}_1\{Q_0\}$. Thus, the *linearized differential equation for the small disturbance* ψ is

(32)
$$\mathscr{L}_0\{\psi\} = a_{\varpi 1} \frac{\partial Q_0}{\partial c_\varpi} + a_{\theta 1} \frac{\partial Q_0}{\partial c_\theta},$$

where $\mathscr{L}_0$ is defined by (27) with $(a_\varpi, a_\theta) = (a_{\varpi 0}, 0)$. With the particular choice of $\Omega(\varpi)$ given by (18), the explicit form of (32) is as follows:

(33)
$$\frac{\partial \psi}{\partial t} + c_\varpi \frac{\partial \psi}{\partial \varpi} + \left(\Omega + \frac{c_\theta}{\varpi}\right) \frac{\partial \psi}{\partial \theta}$$

$$+ \left(2\Omega + \frac{c_\theta}{\varpi}\right) c_\theta \frac{\partial \psi}{\partial c_\varpi} - \left(\frac{\kappa^2}{2\Omega} + \frac{c_\theta}{\varpi}\right) c_\varpi \frac{\partial \psi}{\partial c_\theta} = a_{\varpi 1} \frac{\partial Q_0}{\partial c_\varpi} + a_{\theta 1} \frac{\partial Q_0}{\partial c_\theta}.$$

This is, then, the differential equation serving as the basis of much of our subsequent investigations. After ψ is obtained, the density perturbation $\sigma_1^{(s)}$ produced by the disturbance field $(a_{\varpi 1}, a_{\theta 1})$ may be obtained from the formula

$$(34) \qquad \sigma_1^{(s)} = m_* \int \int \psi \Psi_0 dc_\varpi \, dc_\theta = m_* \int \int \psi e^{-Q_0} dc_\varpi \, dc_\theta.$$

This produces an induced field $(a_{\varpi 1}^{(s)}, a_{\theta 1}^{(s)})$ which forms a part of the *resultant impressed field* $(a_{\varpi 1}, a_{\theta 1})$. If there is no other field than the self-induced field, we are dealing with the problem of a self-maintained *collective mode* of oscillation, whose nature will be discussed in subsequent sections.

3. **Density waves of a spiral form.** Clearly, the variables t and θ are cyclic in the left-hand side of (33). We may thus look for solutions which depend on them in the form $\exp\{i(\omega t - m\theta)\}$, where m is an integer and ω is a complex parameter. We shall see that these solutions lead to *spiral waves*, and here we have the first indication that there could exist, over a stellar disk, density waves of the spiral form which are more or less self-sustained. For convenience of calculation, especially for calculating the joint effect of gas and stars, it is, however, convenient to start with the *resultant* gravitational field, and to consider the response of the stellar disk to such a field with a spiral pattern.

Let us, therefore, consider a gravitational potential given by the real part of

$$(35) \qquad \mathscr{V}_1 = A(\varpi) \exp\{i[\omega t - m\theta + \Phi(\varpi)]\},$$

where $A(\varpi)$ and $\Phi(\varpi)$ are both real. The change of stellar distribution, caused by the imposed gravitational field

$$(36) \qquad a_{\varpi 1} = - \partial \mathscr{V}_1 / \partial \varpi, \qquad a_{\theta 1} = - (1/\varpi)(\partial \mathscr{V}_1 / \partial \theta),$$

can be calculated by (33). It may be expected that ψ is of the form

$$(37) \qquad \psi = \phi(\varpi, c_\varpi, c_\theta) \exp\{i[\omega t - m\theta + \Phi(\varpi)]\},$$

where ϕ varies *slowly* with ϖ, and is in general complex.

All the quantities in our problem now have a *spiral appearance*: the spiral being defined by lines of constant phase of the gravitational potential (35):

$$(38) \qquad m(\theta - \theta_0) = \Phi(\varpi) - \Phi(\varpi_0).$$

The spiral arms are *leading if* $\Phi'(\varpi) > 0$ *and trailing if* $\Phi'(\varpi) < 0$. The circular velocity of the stars is always taken in the counter-clockwise sense.

A crude approximation to the spiral pattern in our galaxy is given by taking

$$(39) \qquad \Phi(\varpi) = -20 \log \varpi.$$

Indeed, for any *smooth* spiral which is relatively tightly wound,[12] we should have

$$(40) \qquad \Phi(\varpi) = \epsilon^{-1} \phi(\varpi) \quad \epsilon \ll 1,$$

where ϵ is a conveniently chosen real parameter, to be identified with the order of magnitude of the inclination of the spiral arms to the circular direction. Thus, a natural method for obtaining simple approximate answers is to adopt an asymptotic series of the type

$$(41) \qquad \phi(\varpi, c_\varpi, c_\theta, \epsilon) = \phi^{(0)}(\varpi, c_\varpi, c_\theta) + \epsilon \phi^{(1)}(\varpi, c_\varpi, c_\theta) + \cdots.$$

Note that the asymptotic procedure is based on the rapidity of the variation of the phase angle $\Phi(\varpi)$ similar to the well-known method of Jeffreys (usually called the WKB method in quantum mechanics).

When the expression (37) is substituted into (33), we obtain the following equation for $\phi(\varpi, c_\varpi, c_\theta)$:

$$
\begin{aligned}
(42) \qquad & c_\varpi \frac{\partial \phi}{\partial \varpi} + \left(2\Omega + \frac{c_\theta}{\varpi}\right) c_\theta \frac{\partial \phi}{\partial c_\varpi} - \left(\frac{\kappa^2}{2\Omega} + \frac{c_\theta}{\varpi}\right) c_\varpi \frac{\partial \phi}{\partial c_\theta} \\
& + i \left[k c_\varpi - m \frac{c_\theta}{\varpi} + (\omega - m\Omega) \right] \phi \\
& = g_\varpi \frac{\partial Q_0}{\partial c_\varpi} + g_\theta \frac{\partial Q_0}{\partial c_\theta},
\end{aligned}
$$

where we have written

$$(43) \qquad k(\varpi) = \Phi'(\varpi),$$

---

[12] Smoothness insures that the higher derivatives of the phase function $\Phi(\varpi)$ are of the same order of magnitude as its first derivative.

so that $k(\varpi) < 0$ for trailing arms, and

(44) $\qquad (a_{\varpi 1}, a_{\theta 1}) = (g_\varpi, g_\theta) \exp\{i[\omega t - m\theta + \Phi(\varpi)]\}.$

Thus,

(45) $\qquad\qquad g_\varpi = -[A'(\varpi) + ik(\varpi)A(\varpi)],$

(46) $\qquad\qquad\qquad g_\theta = imA(\varpi)/\varpi.$

Equation (42) is the partial differential equation for the amplitude function and should be integrated according to the asymptotic scheme (41). We also note that two of its characteristic integrals are those for the original symmetrical distribution [Equations (21) and (22)].

Characteristic integrals. While the angular momentum integral is easily found in the form (21), the energy integral (22) is usually found and used in a form modified through the use of the angular momentum integral. Indeed, for small values of (c_ϖ, c_θ), the "energy integral" takes on the ellipsoidal form

(47) $\qquad\qquad \xi^2 + \eta^2 = \text{function of } \varpi,$

where ξ and η are defined by

(48) $\qquad c_\varpi = \xi(2\varpi\Omega^2/\kappa), \qquad c_\theta = \eta(\varpi\Omega).$

The relation of (47) to the Schwarzschild distribution is obvious. A more careful calculation shows that the energy integral is given in the form

(49) $\qquad \mathscr{E} = \frac{1}{2}\xi^2(1+\eta)^{-\beta} + \int_0^\eta (1+\eta)^{-(\beta+1)}\left(1 + \frac{\eta}{2}\right)\eta\,d\eta,$

for all cases where $\Omega(\varpi)$ varies as a power of ϖ, where

(50) $\qquad \beta = \frac{1}{2}\left(\frac{2\Omega}{\kappa}\right)^2 \frac{\varpi}{\Theta_c}\frac{d\Theta_c}{d\varpi}, \quad \text{and} \quad \Theta_c = \varpi\Omega.$

The integral for angular momentum is

(51) $\qquad\qquad J = \varpi^2\Omega(1+\eta),$

and thus depends essentially on ϖ when η is small; i.e., when the peculiar velocity components are small compared with the circular velocity.

Appendix III

Dynamics of a Gaseous Disk at Zero Pressure

The study of density waves over the whole disk is obviously a very complicated problem. Apart from the attempts of Kalnajs [4] for a stellar disk with finite dispersions all calculations have been made for the case of the gaseous disk at zero pressure or equivalently for a stellar disk with zero dispersion. Toomre [19] considered the possible instability of axially symmetrical disturbances, imposing a condition of *regularity* at the center ($\varpi = 0$) for both the disturbance and the basic distribution of density. He found that short waves are unstable. Rehm [14] calculated the non-symmetrical modes, again imposing the condition of *regularity* at the center. He found that if the basic system has differential

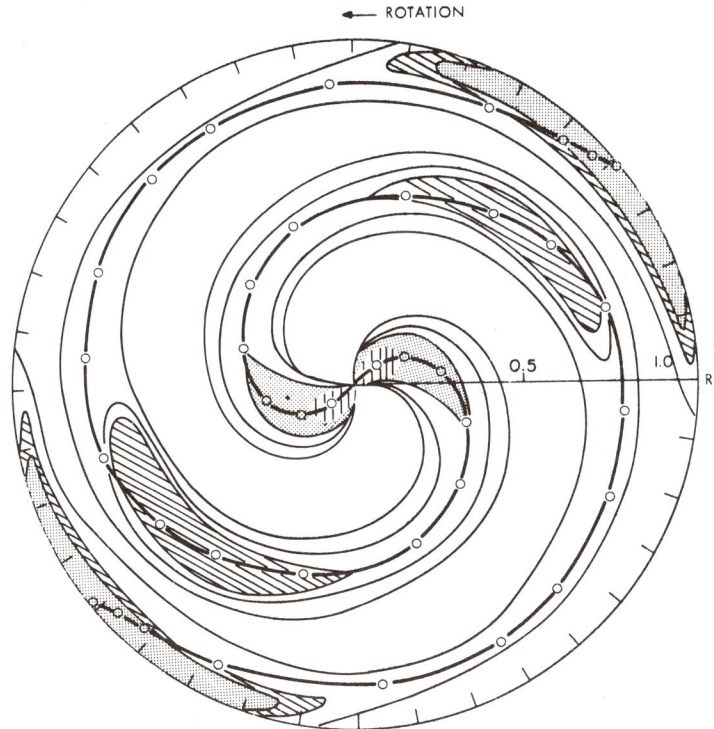

FIGURE A. A density wave pattern (after Rehm [14]).

rotation of the usual type, the amplified spiral patterns are *leading*.[13] Both authors used numerical methods, and it is therefore difficult to conclude whether this is a general feature, but it is presumably so. Still, it would be desirable to examine the problem by the more flexible analytical methods, and to determine whether one can find *trailing patterns* if one allows *singularity* of a vaguely barred shape at the center, as suggested by the continuation of two arms into the center. This was done by Rehm [14] for certain special models studied also by Hunter [2]. Hunter previously considered oscillations over a uniformly rotating disk. He found that the disturbances *regular* at the center show neither leading nor trailing character. He also suggested the consideration of disturbances over a class of differentially rotating disks which permits analytical treatment. Rehm [14] found that the *regular* solutions over these disks with differential rotation again show *leading spiral patterns*. However, he was also able to show the existence of both *trailing and leading spirals if the solution is allowed to have a bar-like singularity at the center*. (These are solutions expressible in terms of series of Legendre functions P_n and Q_n with a few terms of each kind.) Figure A shows a pattern obtained by him. He also found that, in the middle part of the radial range (neither close to the center nor close to the rim of his disk), the pattern formula given by Lin and Shu [5] yields a reasonably good approximation.[14]

The bar-like singularity might be regarded as a *source* for driving the spiral wave. On the other hand, if the disk proper prefers a trailing spiral pattern (as mentioned in §6 of the text for the case of a stellar disk with velocity dispersion), this bar-like singularity could be regarded as a *sink*. After all, if a two-armed pattern is continued right to the center of the disk, some bar-like pattern is almost unavoidable. Rehm's calculations would at least give a crude indication as to how dynamical compatibility could be maintained. In the actual galaxies, the large velocity dispersion of the stars in the three-dimensional nuclear region would be able to adjust to any requirement from the disk proper.

---

[13] It is significant that P. O. Lindblad [7] obtained leading patterns in the early stages of the development in his numerical investigations of the stability of a material disk.

[14] Although Rehm's calculations were made only in the case of uniform rotation, it is safe to conjecture that the result would not be greatly changed by differential rotation.

References

1. S. Chandrasekhar, *Principles of stellar dynamics*, Dover, New York, 1942, 1960.

2. C. Hunter, Monthly Notices Roy. Astronom. Soc. **126** (1963), 299.

3. W. H. Julian and A. Toomre, *On the non-axisymmetric stability and responses of a differentially rotating disk of stars*, Astrophys. J. (to appear).

4. A. J. Kalnajs, Ph. D. Thesis, Department of Astronomy, Harvard University, Cambridge, Mass., 1965.

5. C. C. Lin and F. H. Shu, Astrophys. J. **140** (1964), 646-655.

6. B. Linblad, Stockholm Obs. Ann. **22** (1963), 3-20.

7. P. O. Lindblad, Stockholm Obs. Ann. **21** (1960), 3-73.

8. _____, *Interstellar matter in galaxies*, L. Woltjer, ed., W. A. Benjamin, New York, 1962, pp. 222-233.

9. D. Lynden-Bell, Monthly Notices Roy. Astronom. Soc. **124** (1962), 279.

10. D. Lynden-Bell and P. Goldreich, Monthly Notices Roy. Astronom. Soc. **130** (1965), 97-158.

11. P. Morrison, Rev. Modern Phys. **29** (1957), 235.

12. J. H. Oort, *Interstellar matter in galaxies*, L. Woltjer, ed., W. A. Benjamin, New York, 1962, pp. 3-22, 234-244.

13. K. H. Prendergast, *Interstellar matter in galaxies*, L. Woltjer, ed., W. A. Benjamin, New York, 1962, pp. 217-221.

14. R. G. Rehm, Ph. D. Thesis, Massachusetts Institute of Technology, Cambridge, Mass., 1965.

15. A. Sandage, *The Hubble atlas of galaxies*, Carnegie Institution of Washington, Washington, D. C., 1961.

16. M. Schmidt, B. A. N. **13** (1956), 15.

17. S. Sharpless and O. G. Franz, Publ. Astronom. Soc. Pacific **75** (1963), 219.

18. L. Spitzer, *Physics of fully ionized gases*, Interscience, New York, 1956, pp. 73, 84.

19. A. Toomre, Astrophys. J. **139** (1964), 1217.

20. F. Zwicky, *Morphological astronomy*, Springer-Verlag, Berlin, 1957, pp. 198-201.

MASSACHUSETTS INSTITUTE OF TECHNOLOGY
CAMBRIDGE, MASSACHUSETTS

George Contopoulos

Applications

of the Third Integral

in the Galaxy

1. If we assume that the Galaxy has an axis of symmetry, i.e., its potential is of the form

(1) $$V = V(r, z),$$

then two integrals of motion are the energy

(2) $$H = \tfrac{1}{2}(R^2 + \Theta^2 + Z^2) + V(r, z) = h$$

and the angular momentum

(3) $$C = r\Theta.$$

Let us consider the motion of a star in a meridian plane going through it and through the axis of symmetry of the Galaxy. This is a two-dimensional motion with Hamiltonian

(4) $$H = \tfrac{1}{2}(R^2 + Z^2 + C^2/r^2) + V(r, z) = h$$

that takes place inside the curve of zero velocity,

(5) $$C^2/2r^2 + V(r, z) = h.$$

Equation (4) represents a supersurface in the four-dimensional phase space. We consider the general case when this curve is closed. If the motion is ergodic on this supersurface then the orbit

98

on the plane r, z fills the area inside the curve of zero velocity, i.e., it goes through the neighborhood of every point inside it. If, however, there is a third analytic integral of motion

(6) $$\Phi(r, z, R, Z) = \text{const},$$

then the orbit fills only part of the above area [1]; this was observed in many models of the Galaxy [1], [2], [3], [4], [5], [6], [7], [8], and it seems that it is the general case.

The problem may be stated, in general, as follows: If H is a two-dimensional time-independent Hamiltonian in the coordinate system x, y, (where X, Y are the corresponding momenta), under what conditions is there another time independent isolating integral

(7) $$\Phi(x, y, X, Y) = \text{const}$$

besides the energy? By "isolating" we mean that if Equation (7) is solved for any variable x, y, X, Y, for a given value of h, it gives a finite number of solutions inside the curve of zero velocity.

Any integral of motion Φ satisfies the equation

(8) $$\frac{d\Phi}{dt} = (\Phi, H) = \frac{\partial \Phi}{\partial x}\frac{\partial H}{\partial X} + \frac{\partial \Phi}{\partial y}\frac{\partial H}{\partial Y} - \frac{\partial \Phi}{\partial X}\frac{\partial H}{\partial x} - \frac{\partial \Phi}{\partial Y}\frac{\partial H}{\partial y} = 0.$$

If H is given in the form

(9) $$H = H_0 + \epsilon_1 H_{\epsilon_1} + \epsilon_2 H_{\epsilon_2} + \cdots,$$

where $\epsilon_1, \epsilon_2 \cdots$ are small parameters and

(10) $$H_0 = \tfrac{1}{2}(Ax^2 + By^2 + X^2 + Y^2),$$

we search for integrals Φ of the form

(11) $$\begin{aligned} \Phi_i &= \Phi_{i0} + \epsilon_1 \Phi_{i\epsilon_1} + \epsilon_2 \Phi_{i\epsilon_2} + \cdots + \epsilon_1^2 \Phi_{i\epsilon_1(2)} \\ &\quad + \epsilon_1\epsilon_2 \Phi_{i\epsilon_1\epsilon_2} + \epsilon_2^2 \Phi_{i\epsilon_2(2)} + \cdots. \end{aligned}$$

Equation (8) can be written

(12) $$(\Phi_{i0}, H_0) + \epsilon_1 [(\Phi_{i\epsilon_1}, H_0) + (\Phi_{i0}, H_{\epsilon_1})] + \cdots = 0;$$

hence we may write

(13) $$(\Phi_{i0}, H_0) = 0,$$

(14) $$(\Phi_{i\epsilon_1}, H_0) = -(\Phi_{i0}, H_{\epsilon_1}),$$

etc.

Equation (13) shows that Φ_{i0} is an integral of the unperturbed Hamiltonian (10) (when $\epsilon_1 = \epsilon_2 = \cdots = 0$).

This Hamiltonian is very simple. It is known that if $A^{1/2}/B^{1/2}$ is irrational, then the only isolating integrals are

(15) $\Phi_{10} = \tfrac{1}{2}\,(Ax^2 + X^2) = \text{const}$

and

(16) $\Phi_{20} = \tfrac{1}{2}\,(By^2 + Y^2) = \text{const}.$

After Φ_{i0} are found we can find $\Phi_{i\epsilon_1}$ from Equation (14). This is a linear partial differential equation with the second member a known function. The corresponding characteristic system is

(17) $$\frac{dx}{X} = \frac{dX}{-Ax} = \frac{dy}{Y} = \frac{dY}{-By} = \frac{d\Phi_{i\epsilon_1}}{-(\Phi_{i0}, H_{\epsilon_1})}.$$

If we introduce the auxiliary variable t through $dt = dx/X$, we find

(18)
$$A^{1/2}x = (2\Phi_{10})^{1/2}\sin A^{1/2}(t - t_0), \quad B^{1/2}y = (2\Phi_{20})^{1/2}\sin B^{1/2}t,$$
$$X = (2\Phi_{10})^{1/2}\cos A^{1/2}(t - t_0), \qquad Y = (2\Phi_{20})^{1/2}\cos B^{1/2}t.$$

Then we find

(19) $$\Phi_{i\epsilon_1} = -\int (\Phi_{i0}, H_{\epsilon_1})\, dt,$$

where x, X, y, Y have been replaced by their values (18). The integration gives $\Phi_{i\epsilon_1}$ as a polynomial in the quantities (18), unless $-(\Phi_{i0}, H_{\epsilon_1})$ contains a constant term q, which gives a secular term qt in $\Phi_{i\epsilon_1}$. It can be proven [1] that no such terms appear when $A^{1/2}/B^{1/2}$ is irrational. If $A^{1/2}/B^{1/2}$ is rational there are always such terms, but these can be removed by a convenient combination of the integrals of motion [9].

In the case of the field near the sun we have approximately

(20) $$H = \frac{1}{2}\,(Ax^2 + By^2 + X^2 + Y^2) - \epsilon xy^2 - \frac{\epsilon' x^3}{3} = h,$$

with

$$x = r - r_0, \quad y = z, \quad X = R, \quad Y = Z, \quad A = 0.076(10^7\,\text{yr})^{-2}$$
$$B = 0.55(10^7\,\text{yr})^{-2}, \quad \epsilon = 0.206\,\text{kpc}^{-1}(10^7\,\text{yr})^{-2},$$
$$\epsilon' = 0.052\,\text{kpc}^{-1}(10^7\,\text{yr})^{-2}.$$

The third integral is then

$$\Phi_1 = \frac{1}{2}(Ax^2 + X^2) + \frac{\epsilon}{4B - A}[(A - 2B)xy^2 - 2xY^2 + 2XyY]$$
(21)
$$+ \frac{\epsilon' x^3}{3} + \cdots.$$

The convergence of the third integral is unknown. However the following general theorem has been proved [10]: If H is given as a series, then we can find another series $\overline{H}$, coinciding with the given one up to the terms of any degree, given a priori, which has convergent integrals of the form (11).

Therefore if we remark that the higher order terms of the galactic potential are not known we may find a model of the Galaxy which is separable for small x and y, i.e. near the sun.

Van de Hulst [11] has given a potential for the neighborhood of the sun agreeing with the potential of the Hamiltonian (20) up to the terms of 4th degree, which is separable in elliptical coordinates; i.e., if we transform the variables x, y into elliptical coordinates the Hamiltonian is the sum of two parts, each of which contains only one pair of conjugate variables.

Hori [3] gave a more general separable model of the Galaxy. Orbits in nonseparable models have been given by Ollongren and Torgård [4], [5], [6], by Perek and Peterson [7] and by Hayli [8].

In the general case of separable potentials, when the ratio of the frequencies along the two main directions in phase space (corresponding to $A^{1/2}/B^{1/2}$ of the unperturbed case (10)) is irrational, all orbits are "box orbits," i.e., they fill curvilinear parallelograms whose apices are on the curve of zero velocity (Figure 1). If, however, the ratio of the frequencies is rational all orbits are periodic.

In the case of nonseparable potentials we may define frequencies, corresponding to $A^{1/2}$, $B^{1/2}$, which are variable. Therefore for some initial conditions the ratio of the frequencies may become a rational number and then we have a periodic orbit. For initial conditions near the periodic orbit we may have "tube" orbits, i.e., elongated orbits that never go far away from the periodic orbit (Figure 2); then the periodic orbit is stable. Such "tube" orbits were found by Torgård [6] and Ollongren [12]. An explanation of these orbits by means of the third integral was given lately [13].

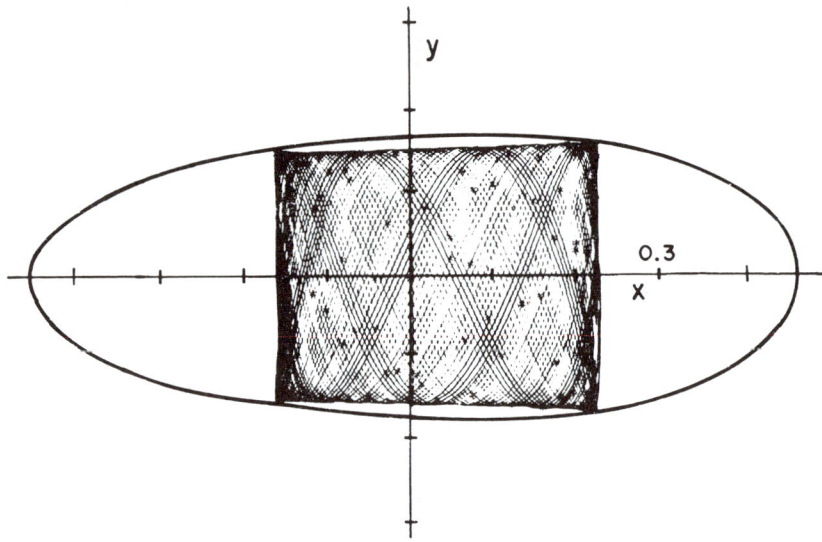

FIGURE 1. Box-orbit in the potential
$V = (1/2)(Ax^2 + By^2) - \epsilon xy^2$ with $A = 0.076$, $B = 0.55$,
$\epsilon = 0.206$ and $x_0 = y_0 = 0$, $X_0 = 0.0512$, $Y_0 = 0.1126$.
The solid curve is the curve of zero velocity.

In the models of Ollongren, Perek, etc., if the energy of a star is very large (but smaller than the energy of escape), its orbit seems not to lie any more on an analytic supersurface (6). If an integral of motion (7) gives infinitely many values for one variable, when the other variables are fixed, it is called nonisolating. This case can be subdivided into two: if the solutions do not fill the whole space available but only part of it we have the quasi-isolating case; if the solutions fill the whole available space we have the ergodic case (quasi-ergodic in older terminology).

Moser [14] and Arnol'd [15] have proved that the quasi-isolating case is the most general one. In fact they proved, under some very general conditions, that there exist integral surfaces in phase space (the intersections of H and Φ), which form a set of positive measure, but do not form necessarily a continuous set. There may be exceptions of positive measure. However any point that is between two integral surfaces remains always between them; the corresponding integral may give infinitely many values, which are between two limits.

The transition from the isolating to the quasi-isolating case

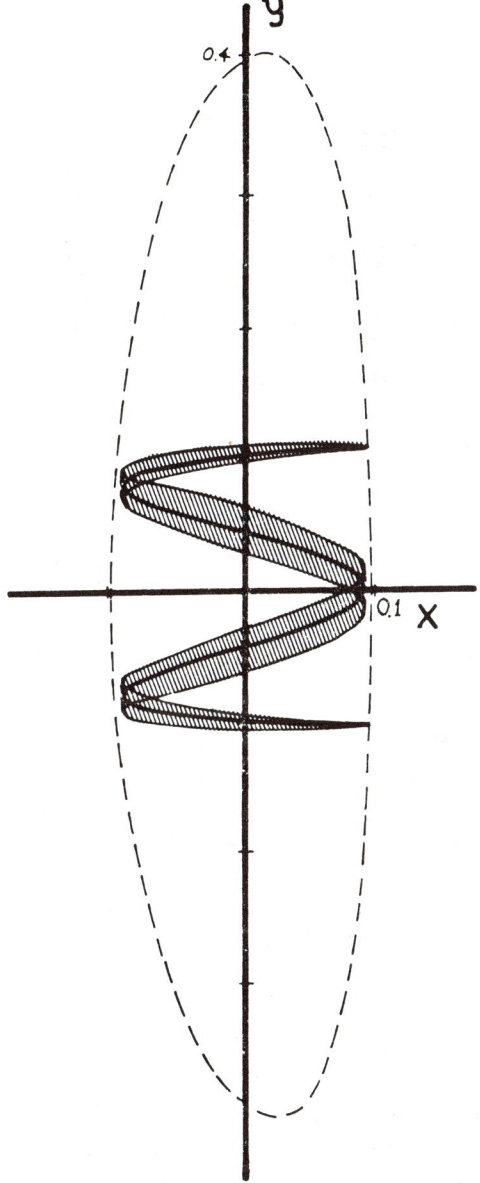

FIGURE 2. Tube-orbit in the same potential with $A = 1.6$, $B = 0.1$, $\epsilon = 0.2$ and $x_0 = 0.0936$, $y_0 = X_0 = 0$, $Y_0 = 0.0358$. The dashed curve is the curve of zero velocity.

seems to be abrupt. This is shown in Figure 3(a), (b), which compares the results found by Hénon and Heiles [16] and Barbanis [17] as regards the proportions of isolating orbits in a given field of the form (20), with $A = B = \epsilon$.

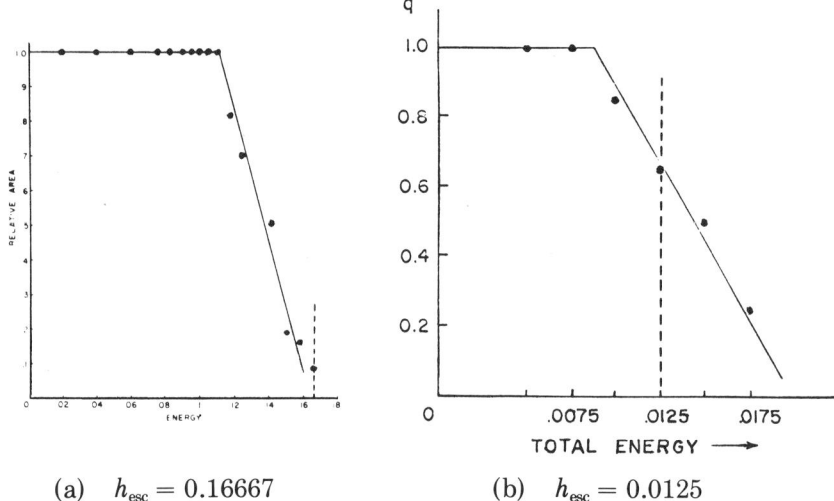

(a) $h_{esc} = 0.16667$ (b) $h_{esc} = 0.0125$

FIGURE 3. Proportion of isolating orbits as a function
of the energy. The dashed line gives the energy of escape.

This proportion is found as follows. If Φ is an isolating integral, then the projection of the intersection of the hypersurfaces $H = h$ and $\Phi = \text{const}$ on a three-dimensional space (say, xyX) is an isolating surface. If we intersect it by a plane (say, $y = 0$) it gives a simple (isolating) curve. Empirically such a curve is found if we mark the successive points of intersection of an orbit by the x-axis and the corresponding X-velocities, and plot the points (x, X). If all such points lie on a simple curve we have an indication that the orbit lies on an isolating integral surface. If the points (x, X) have a noticeable spreading we have the quasi-isolating case, and if they fill the whole available space we have the ergodic case.

In all cases we can measure empirically the proportion of the area covered by well-defined curves and plot it versus the value of the energy h.

Hénon and Heiles take $A = \epsilon' = 1$, $\epsilon = -1$ while Barbanis takes $A = \epsilon = 0.1$, $\epsilon' = 0$. In both papers all the orbits seem to be isolating up to a certain value of the energy. If h increases beyond that value part of the orbits become quasi-isolating, and finally ergodic. In Hénon and Heiles' case the region of isolating orbits becomes zero at the energy of escape (when the curve of zero velocity opens and the moving point may go to infinity). In Barbanis' case even beyond the escape energy part of the orbits do not escape, but remain near a stable periodic orbit. Part of these orbits give quite good invariant curves, even for energies up to 1.5 times the escape energy.

In another, unpublished, case, Hénon has found that all orbits seem to be ergodic up to the energy of escape. The problem of finding why and how an integral begins to become quasi-isolating and ergodic is not yet solved.

Similar results for the galactic potential were found by Perek and Peterson [7], and Hayli [8].

According to Perek and Peterson the third integral ceases to be nearly isolating for velocities at the center of the Galaxy of the order of 400 km/sec. It seems that for all ordinary stars, moving inside the main body of the Galaxy, there is a very accurate third integral. For larger orbits, e.g., for the orbits of globular clusters it seems that the third integral becomes ergodic. The corresponding invariant curves are no more simple curves but fill the whole available area.

On the other hand all orbits in Hori's model [3] are boxes, and no such effects appear.

Therefore in order to see if the results of Perek, Peterson, and Hayli represent real effects in our Galaxy, one has first to make the following check: Construct a set of separable potentials representing approximately the galactic field, as we know it up to the terms of third, or fourth, or higher order. The higher order terms of these potentials will not influence the inner parts of the galactic model (or the parts near the sun in the case of a model of the form (20)). However the outer parts will be influenced very much by them. If the models differ from the observed galactic potential more than it is permitted by the inaccuracies of observations, then we may say that the real potential is definitely nonseparable, and quasi-isolating or ergodic orbits do exist. The same refers to the tube orbits which exist in some but not all models [3], [5], [6].

If a separable potential is within the limits set by observation then all the above effects may not be real. In any case it would be useful to see if slightly different potentials give the same results as regards the position of the periodic, tube, and quasi-isolating orbits. It seems that the periodic orbits are somehow connected with the appearance of the quasi-isolating case. In fact our numerical experience indicates that the first nonisolating orbits appear near the unstable periodic orbits [17].

The formulae that give the higher order terms of the third integral are known [1], but the actual calculation of these terms becomes soon prohibitively long.

We have constructed recently a computer program to find higher order terms of the third integral for any potential of the form (9). An IBM 7094 computer can calculate the terms up to the 10th degree in the variables (for only one term ϵH_ϵ) in 2.5 minutes. But the time needed to go to the 14th degree is of the order of 40 minutes. The values found in the specific case that $H_\epsilon = - xy^2$, will be given in a future paper. The number of terms of different degrees are: (2) 2; (3) 3; (4) 8; (5) 11; (6) 20; (7) 26; (8) 40; (9) 50; (10) 70; (11) 85; (12) 112; (13) 133; (4) 168.

TABLE I. Constancy of the Third Integral

	max	min	D	$2D/(\max + \min)$
Φ_{10}	.188	.101	.087	0.602
$+ \epsilon \Phi_{1\epsilon}$	.139	.129	.010	0.075
$+ \epsilon^2 \Phi_{1\epsilon(2)}$	.1337	.1322	.0015	0.011
$+ \epsilon^3 \Phi_{1\epsilon(3)}$	.1326	.1323	.0003	0.002
$+ \epsilon^4 \Phi_{1\epsilon(4)}$	.13247	.13241	.00006	0.0005
$+ \epsilon^5 \Phi_{1\epsilon(5)}$	.132448	.132424	.000024	0.0002
$+ \epsilon^6 \Phi_{1\epsilon(6)}$	.132444	.132422	.000022	0.00017
$+ \epsilon^7 \Phi_{1\epsilon(7)}$	.132444	.132430	.000014	0.00011
$+ \epsilon^8 \Phi_{1\epsilon(8)}$	.132439	.132431	.000008	0.00006
$+ \epsilon^9 \Phi_{1\epsilon(9)}$	.132438	.132434	.000004	0.00003
$+ \epsilon^{10} \Phi_{1\epsilon(10)}$	.132437	.132435	.000002	0.000015
$+ \epsilon^{11} \Phi_{1\epsilon(11)}$	.1324365	.1324358	.0000007	0.000005
$+ \epsilon^{12} \Phi_{1\epsilon(12)}$	.1324361	.1324358	.0000003	0.000002

Units $10^{-2} \text{kpc}^2 (10^7 \text{yr.})^{-2}$

It is remarkable that as higher order terms are included in the expansion (11) the integral is better conserved if the perturbation is small. In order to give the order of magnitude of this effect a number of orbits were calculated and the value of the third integral in different degrees of approximation was found in many points along the orbit.

Table I gives the maximum and minimum values of the third integral (21) for $\epsilon' = 0$ (the other constants are the same as above) truncated after the term $\Phi_{10}, \epsilon\Phi_{1\epsilon}, \cdots, \epsilon^{12}\phi_{1\epsilon(12)}$, for an orbit with initial conditions $x_0 = y_0 = 0$, $X_0 = 0.0512$, $Y_0 = 0.1126$.

The maximum deviation $D = \max - \min$ and the maximum relative deviation $2D/(\max + \min)$ are also given in this Table. It is seen that the deviations decrease every time higher order terms are added. When terms up to the 14th degree (12th order) are included almost 7 significant figures of the third integral are constant.

The energy constant was conserved up to the last printed figure (eighth).

2. We have discussed the application of the third integral in explaining the forms of the galactic orbits in meridian planes following the stars in their motions. Another application refers to the plane galactic orbits if the galactic potential is not exactly axisymmetric, but has a plane of symmetry. In this case the angular momentum is no more conserved. It seems, however, that one can find another integral, which is isolating or quasi-isolating in most cases.

If in the Hamiltonian (9) we assume $A = B$ we have, in zero-order approximation, a spherical homogeneous galaxy. We have considered in detail [18] the case where a perturbation of the form $-\epsilon xy^2$ makes the Galaxy asymmetric. It is found that if $A = B = 0.1$, $h = 0.00765$, then for $\epsilon = 0.05$ the maximum asymmetry is 20%, and for $\epsilon = 0.1$ it is about 60%.

We find that the orbits are of three types:

A. Most of the orbits are of type A (Figure 4). These orbits have some similarity to the box orbits of the general irrational case, but there are also some differences. The boundary to the left and right has angular points on the x-axis and it can be continued inwards, forming two inner triangles. The inner arcs of the boundary are envelopes of sets of arcs of the orbit.

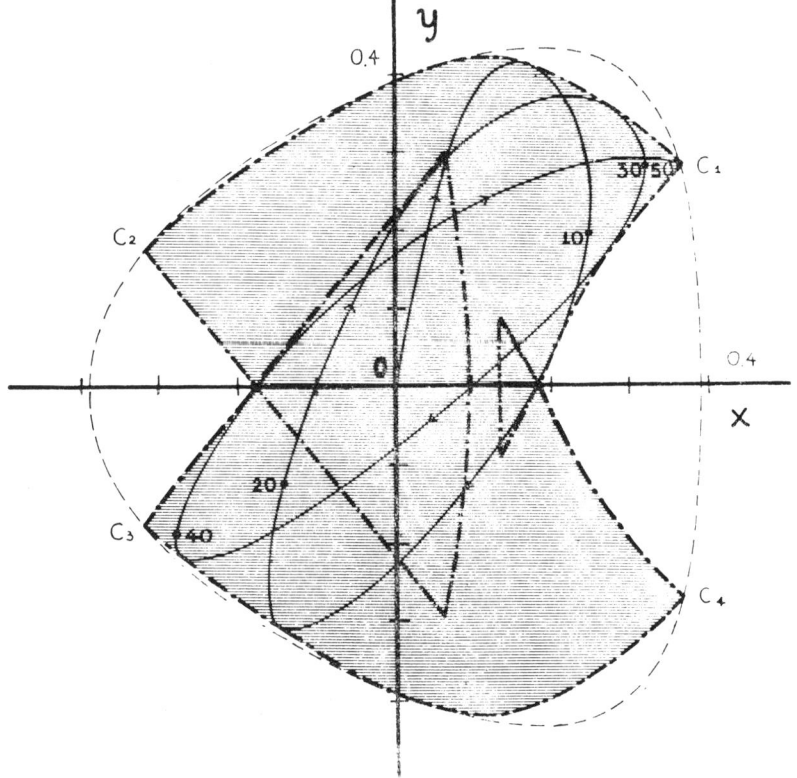

FIGURE 4. A-type orbit: $A = B = \epsilon = 0.1$, $x_0 = y_0 = 0$,
$X_0 = 0.026$, $Y_0 = 0.12093$. The dashed curve is the
curve of zero velocity.

B. The B-type orbits are tube orbits (Figure 5), i.e., elongated orbits surrounding two stable straight line periodic orbits.

C. The C-type orbits form rings around the origin (Figure 6). They do not reach the curve of zero velocity, but surround another stable periodic orbit.

There are two more periodic orbits, the axis $y = Y = 0$, and another one near the y-axis, but they are unstable. In most cases, an appreciable exchange of energy between the two degrees of freedom is observed.

The form of the third integral in this case is rather different from the usual one. Its zero order terms are of fourth degree

(22)
$$\phi_0 = X^2Y^2 - Ax^2Y^2 - AX^2y^2 + A^2x^2y^2 + 4AxXyY$$
$$- (Ax^2 + X^2)^2/3 + (Ay^2 + Y^2)^2/12.$$

If we consider this case as representing a rough model of a galaxy with a distortion of the order 20% we find the following:

(a) Only a few orbits of stars ejected perpendicularly to the x-axis are nearly circles or ellipses (type C orbits). This happens only if the initial point is far from the center and near the curve of zero velocity corresponding to a given energy h.

Orbits of stars ejected perpendicularly nearer to the center (among others, orbits with circular velocity) are type-A orbits, i.e., they gradually become more elongated until they go through the central region of the Galaxy, and then reverse the sense of rotation.

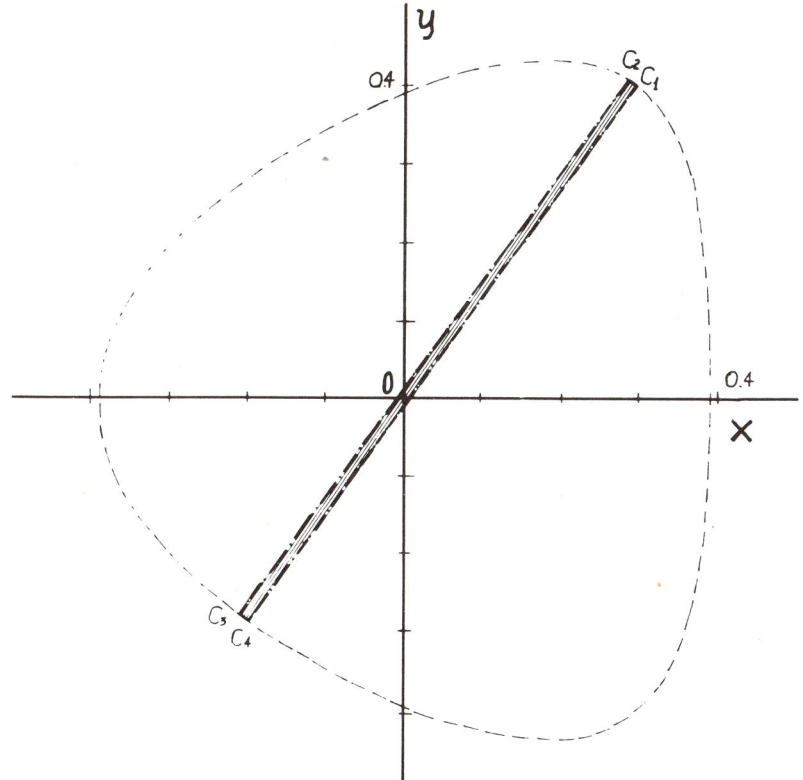

FIGURE 5. B-type orbit: $A = B = \epsilon = 0.1$, $x_0 = y_0 = 0$,
$X_0 = 0.07$, $Y_0 = 0.10198$.

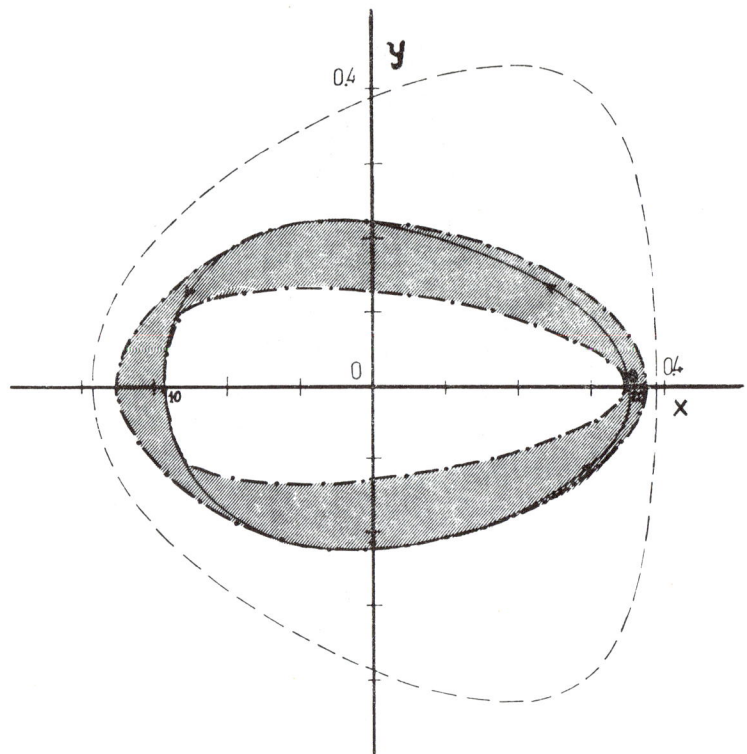

FIGURE 6. C-type orbit: $A = B = \epsilon = 0.1$,
$x_0 = 0.35, y_0 = X_0 = 0, Y_0 = 0.055227$.

E.g., if a star starts at 10 kpc from the center with a circular velocity 250 km/sec, its orbit becomes almost rectilinear after about 10×10^9 yr and becomes circular again, in the opposite direction, after about 20×10^9 yr.

(b) Orbits ejected from the central part of the Galaxy become usually A-type orbits, i.e., they become almost circular or elliptical after some time, except if they are near the periodic orbits of type B.

When the distortion of the Galaxy reaches large values, then most orbits may escape to infinity; in cases near the escape case the orbits become quasi-isolating [17]. For moderately distorted galaxies, however, it seems that the orbits are isolating or very nearly isolating.

It is of special interest to extend such calculations to other models of the Galaxy, especially spiral models. Such a work has

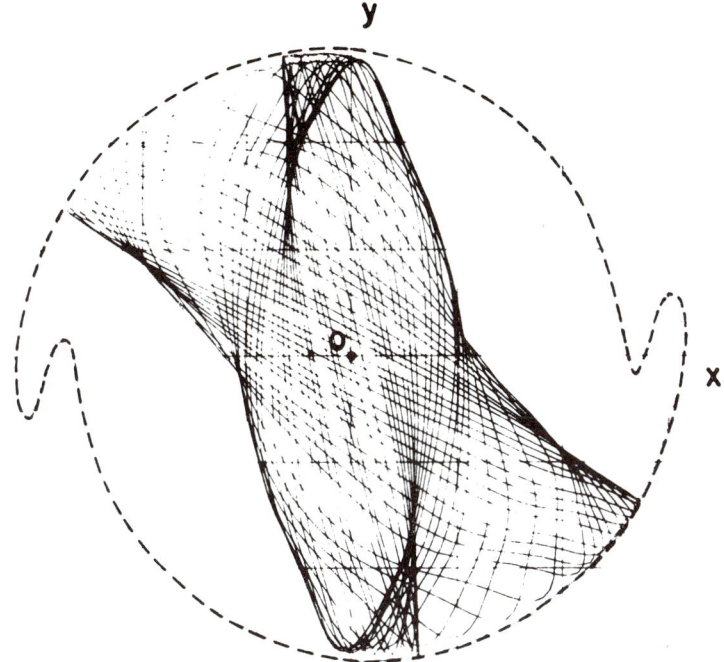

FIGURE 7. Orbit in the potential
$V = (A/2)(x^2 + y^2)(1 + \epsilon \cos(2\theta - \beta r))$ with $A = 10$, $\epsilon = 0.2$,
$\beta = 1$. The initial point is $x_0 = 0.5$, $y_0 = 0$, and it is ejected
perpendicularly to the x-axis. The energy $h = 1$, and the
corresponding curve of zero velocity is a dashed line.

begun recently by Barbanis. I will mention here only a few pre-
liminary results.

As a first step a simple model of a spiral galaxy was considered
with a potential of the form

(23) $\qquad V = (A/2)(x^2 + y^2)\{1 + \epsilon \cos(2\theta - \beta r)\}$,

where r, θ are polar coordinates. In zero order this represents a
homogeneous spherical galaxy, and the spirals

$$2\theta = \beta r$$

are nonrotating.

In the examples considered, the density in the spirals is 20%
larger than the mean density.

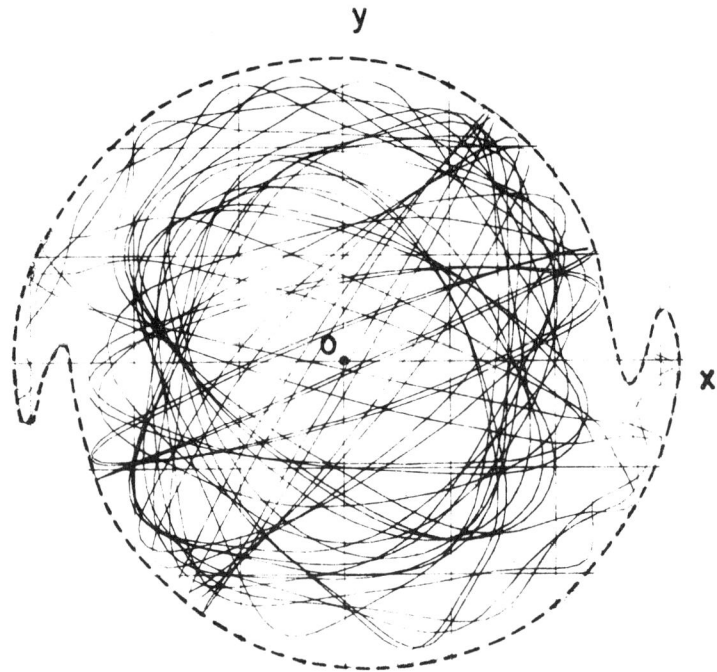

FIGURE 8. Orbit in the same potential and with
the same energy, but with initial point at $x_0 = 1$.

The curve of zero velocity for $h = 1$ is given as a dashed line in
Figures 7 and 8. The orbits are ejected perpendicularly to the
x-axis.

It seems that only for some initial conditions the orbits are rings
around the origin. In Figure 7 the orbit seems to be a broad tube
orbit, but more complicated than those of the case discussed above.
The moving point rotates initially counterclockwise, but gradually
the orbit becomes very elongated and then reverses the direction
of rotation.

In Figure 8 it seems that the orbit is ergodic and fills all the
space inside the curve of zero velocity. Maybe in irregular potentials
the ergodic case is more general than in smooth potentials [23].

The invariant curve in the first case is well defined; it indicates
that the third integral is isolating or very nearly isolating. In the
last case the points (x, X) fill most of the available area, but there

is a relatively small empty space. The orbit either is ergodic or very nearly ergodic.

Further results of this type are expected with much interest.

3. Another application of the third integral in the Galaxy is the explanation of the three-axial form of the velocity ellipsoid.

If the distribution function in the Galaxy is a function of only the energy and the angular momentum

(24) $$f = f(H, C)$$

then the velocity ellipsoid must have symmetry with respect to the Θ-axis (the axis perpendicular to the direction to the center in the plane of symmetry). This is because in f the velocities R and Z appear always through the combination $R^2 + Z^2$. It is known, however, that the observed Z-axis of the velocity ellipsoid is much shorter than R, being nearly equal to the Θ-axis.

If there is a third integral of motion we may take

(25) $$f = f(H, C, \Phi),$$

where $\Phi = R^2/2 + \cdots$, therefore there is no more coupling between R and Z.

Barbanis [19] has taken for the distribution of velocities in the plane R, Z the formula

(26) $$dN = A^{-2kH-2l\Phi} dR dZ,$$

where dN is the number of stars with velocities between R and $R + dR$, Z and $Z + dZ$; A, k, l are constants, and Φ is the third integral truncated after the terms of first degree in ϵ. He found that the observed distribution of stellar velocities is described satisfactorily if we take $k = 19.7 \, \text{kpc}^{-2} (10^7 \, \text{yr})^2$, $l = 14.1 \, \text{kpc}^{-2} (10^7 \, \text{yr})^2$; therefore the role of the third integral is of the same order as that of the energy.

4. A more important application of the third integral refers to the construction of galactic models. The distribution function f, the density ρ, and the potential V of the Galaxy are connected by the formulae

(27) $$\rho = \int \int \int f \, dX \, dY \, dZ,$$

(28) $$\Delta V = 4\pi G \rho,$$

and

(29) $\quad \dfrac{\partial f}{\partial t} + \dfrac{\partial f}{\partial x}\, X + \dfrac{\partial f}{\partial y}\, Y + \dfrac{\partial f}{\partial z}\, Z - \dfrac{\partial f}{\partial X}\dfrac{\partial V}{\partial x} - \dfrac{\partial f}{\partial Y}\dfrac{\partial V}{\partial y} - \dfrac{\partial f}{\partial Z}\dfrac{\partial V}{\partial z} = 0.$

If the potential V is given then ρ can be found from Poisson's Equation (28), and the solution of Liouville's Equation (29) is $f =$ a function of the integrals of motion. Equation (27) is a restriction on the form of the function f.

Lynden-Bell [20] has given explicitly the solution of Equation (27) in axisymmetric models when f is a function of the first two integrals of motion, and ρ is given as a function of the potential V and the distance r. The solution is not unique. If f is one solution, the general solution of Equation (27) is $f + \Delta f$, where Δf is any function of H and C satisfying the conditions

$$\Delta f(H, C) = - \Delta f(H, - C)$$

and $f + \Delta f \geqq 0$ everywhere.

If f depends also on the third integral Φ there is one more degree of arbitrariness. However, there may be some extra conditions that specify the solutions.

We know, e.g., that the ellipsoidal theory of stellar velocities is valid approximately only for small velocities. For large velocities it is not even approximately correct [21]. Fricke has given a general form of the distribution function $f(H, C)$ to account for the distribution of velocities in the plane of symmetry of the Galaxy.

If, more generally, we assume that V and ρ are functions of x, y, z, and f a function of H, C, Φ, then by solving Equation (27) we can find f.

The construction of some models of this type would be useful in understanding the structure and dynamics of the galactic system.

One problem remains to be solved empirically in this connection. Lynden-Bell [22] pointed out that only isolating integrals should be arguments of the distribution function f. This justifies the omission of all integrals beyond the third, because the other integrals are probably ergodic. But in the case of the third integral the problem is different. If the third integral is isolating it should be included in f. If it is quasi-isolating, we may use a truncated function $\overline{\Phi}$, which is only approximately constant. Then Liouville's equation is only approximately satisfied. However, if the error is always small, it is of the same nature as the numerical errors of calculation, and we can solve Equation (27) to give f within a certain degree of uncertainty. If the uncertainty is not large the solution may be accepted. In any case V cannot be known quite accurately, therefore an uncertainty as regards ρ and f is also permissible. The degree of uncertainty has to be found empirically. Such a study will show how useful the quasi-isolating integrals are in constructing models of stellar systems.

References

1. G. Contopoulos, Z. Astrophys. **49** (1960), 273.
2. ———, Stockholms Obs. Ann. No. 5, **20** (1958).
3. G. Hori, Publ. Astr. Soc. Japan **14** (1962), 353; Proc. Internat. Astronom. Union Sympos. No. 25 (1966).
4. A. Ollongren, Bull. Astronom. Inst. Netherlands **16** (1962), 241.
5. ———, Proc. Internat. Astronom. Union Sympos. No. 25 (1966).
6. I. Torgård, Nuffic International Summer Course in Science, Part X (1960).
7. L. Perek and D. M. Peterson, Proc. Internat. Astronom. Union Sympos. No. 25 (1966).
8. A. Hayli, Ann. Astrophys. **28** (1965), 49.
9. G. Contopoulos, Astronom. J. **68** (1963), 763.
10. ———, Astrophys. J. **138** (1963), 1297.
11. H. C. van de Hulst, Bull. Astronom. Inst. Netherlands **16** (1962), 235.
12. A. Ollongren, Ann. Rev. Astronom. Astrophys. **3** (1965), 113.
13. G. Contopoulos, Astronom. J. **70** (1965), 526.
14. J. Moser, Nachr. Akad. Wiss. Göttingen Math.-Phys. Kl. 1 (1962); *Nonlinear Problems*, R. E. Langer, ed., Univ. Wisconsin Press, Madison, Wis., 1963.
15. V. I. Arnol'd, Uspehi Mat. Nauk **18** (1963), 91. (English translation by Air Force Systems Command).
16. M. Hénon and C. Heiles, Astronom. J. **69** (1964), 73.
17. B. Barbanis, Astronom. J. **71** (1965), 415.
18. G. Contopoulos and M. Moutsoular, Astronom. J. **70** (1965), 817.
19. B. Barbanis, Z. Astrophys. **56** (1962), 56.
20. D. Lynden-Bell, Monthly Notices **123** (1962), 447.
21. W. Fricke, Astronom. Nachr. **280** (1951), 125; Naturwissenschaften **38** (1951), 438.
22. D. Lynden-Bell, Monthly Notices **124** (1962), 1.
23. G. Contopoulos and L. Woltjer, Astrophys. J. **140** (1964), 1106.

GODDARD INSTITUTE FOR SPACE STUDIES, NASA
NEW YORK, NEW YORK

Ivan R. King

The Dynamics of Star Clusters

This lecture deals with a topic in the dynamics of stellar systems. Stellar dynamics is a sort of n-body problem, except that we no longer try to follow each of the bodies individually; instead we treat them statistically and deal with distributions of density and velocity. In doing so we idealize the stellar system as if it had infinitely many particles, so that statistical fluctuations can be ignored. In a real stellar system, of course, this will introduce some error; but the needed corrections are small enough that they can be ignored for the moment and added later as an afterthought.

In a stellar system, then, we have at every point a star density and a distribution of velocities. This can be thought of as a "stellar gas"—a velocity (mean velocity, that is) and a pressure (distribution of random motions) at every point. The star gas differs from an ordinary gas, though, in that collisions arc almost infinitely rare. As a result, the random velocities need not be isotropic; mathematically the scalar pressure is replaced by a pressure tensor, and physically there are no viscous forces. Each individual star follows an orbit in the (smoothed-out) gravitational field of all the others—although it is usually more profitable scientifically to look at velocity distributions rather than individual orbits.

116

Stellar systems can be classified dynamically in several ways. First, it makes a big difference whether the system is all stars or has a sizable admixture of gas. As stated in Dr. Lin's paper in these Lectures, the gas behaves rather differently and can have a serious effect on the behavior of the stars. I am going to talk about the simpler systems in which the stars have only each other to worry about. A second big difference is whether or not the system is rapidly rotating. At the one extreme are flattened disc systems in which most of the motion is in systematic rotational velocities and the random motions are relatively small. At the other extreme are the systems I am going to talk about, in which the motions are almost completely of a random sort and rotation adds only a small effect—hydro-dynamically speaking, the internal energy is all pressure and little or no velocity. In fact, I am going to make the problem as simple as possible by restricting the discussion to spherically symmetrical systems, with no net rotation at all.

Astronomically we recognize three types of round stellar systems. Although the distinction is basically generic rather than dynamical, it is a convenient distinction to retain in the present discussion, because the differing numbers of stars and overall sizes do make a difference to the dynamics. Individual sizes range widely within each class, but typical figures are:

Type	Number of stars	Radius
Open cluster	10^3	10 parsecs
Globular cluster	10^5	100
Spherical galaxy	10^{10}	10,000

When we observe these stellar systems, they turn out to have very similar star distributions. This is not too easy to see among open clusters, where statistical fluctuations in star numbers often make a cluster look ragged rather than regular; but globular clusters are individually rather smooth and collectively quite similar. Spherical galaxies (or elliptical galaxies, as we more commonly call them in recognition of the fact that galaxies in the real world are not all perfectly round and rotation-free) have an extremely smooth distribution of light (most are too far away for us to see the individual stars), which builds up to a central brightness peak that is very like that in a globular cluster but often looks sharper because of the distance of the galaxy from us.

I speak only of density distributions. We would certainly be pleased to observe the velocity distributions, but they are unfortunately beyond our reach. The best we can hope to observe is the overall dispersion of velocities in such a stellar system—a quantity that only serves as a scale factor.

All these stellar systems have much in common dynamically; to illustrate their characteristics I shall concentrate on the globular clusters, indicating in a few side remarks how the open clusters and elliptical galaxies differ.

The basic dynamics of globular clusters can be clearly understood in terms of the relative rates of different processes—or equivalently, in terms of the characteristic times of those processes. The fastest process is mixing. Each star moves as an individual in the smooth gravitational field of the whole assembly, and neighbors soon go their separate ways. Thus there can be no phase-coherent phenomena such as overall pulsation of the cluster. The time scale for phase mixing—and consequent smoothing out of the density distribution—is the orbital period of a star, which in a typical globular cluster is about 10^6 years. This mixing time is analogous to the free oscillation time of the system—except that any systematic oscillation would be rapidly damped out.

The second basic process is relaxation. This is the only point at which we need to consider the fine-grained nature of the density distribution. What happens is that individual stars occasionally encounter each other at a small enough distance that their orbits in the cluster are changed. Such an encounter can be idealized as a hyperbolic two-body orbit whose net effect, in a fixed coordinate system, is a transfer of energy from one star to the other. Since the stars are distributed randomly, the circumstances of the encounters are random and so are the energy exchanges. Statistical mechanics tells us that the tendency will then be toward a Maxwellian velocity distribution, with equipartition of energy between stellar groups of different individual mass. As we will see, this tendency never reaches full fruition; but it operates strongly nevertheless, and the changes that do take place give us our second characteristic time: the relaxation time. At the center of a typical globular cluster it is about 10^8 years.

The third basic process is dynamical evolution. It is a consequence of relaxation, which tries to give the velocity distribution a Gaussian shape. A cluster of finite mass has, however, a finite escape velocity; so a Gaussian velocity distribution would include some stars that will immediately escape. Relaxation toward a Gaussian velocity distribution will thus cause a continuing loss of stars. Detailed analyses of the process, which I shall describe later in this lecture, give a loss rate of rather less than 1 per cent per relaxation time in a typical globular cluster. We may thus describe the dynamical evolution by a characteristic time that is several hundred times the relaxation time—or, in a typical globular cluster, several times 10^{10} years.

Along with these three characteristic dynamical time scales, which exist in general in what we might call the abstract problem of cluster dynamics, there is another time scale that we should be aware of. It belongs to the specific astronomical problem. From studying the physical evolution of individual stars we know that the globular clusters that we observe are about 10^{10} years old. The elliptical galaxies also appear to have this age. Open clusters, on the other hand, range from as young as 10^{6} years up to the maximum age of 10^{10} years.

Comparison of the age with the characteristic dynamical times shows which processes are significant for each type of system. Thus we may expect globular clusters to be completely mixed and thoroughly relaxed, but their dynamical evolution is barely under way. In elliptical galaxies all the dynamical time scales are about 100 times as long; hence they are mixed but unrelaxed and unevolved. A typical open cluster, on the other hand, has for its three characteristic dynamical time scales 10^{6}, 10^{7}, and 10^{9} years, respectively. Hence all open clusters are mixed, most are relaxed, and some are highly evolved. In fact, many of the open clusters formed in the Milky Way are already dead, in the sense that their stars have nearly all been lost.

I should like to discuss the dynamics of star clusters in two stages. First we will look at the gross dynamics, dealing with overall averages for the cluster and aiming only at determining the characteristic time scales. Then we will look at the problem in more detail and attempt to determine the distributions of velocity and density that are implied by the relaxation process.

The first important relation for any self-gravitating system is the virial theorem. It says [1, p. 200] that

(1) $2T = -V,$

where T is the total internal kinetic energy and V is the gravitational potential energy of the system. We can write

(2) $2T = M\langle v^2\rangle,$

where M is the total mass and $\langle v^2\rangle$ is the mass average of the square of the velocity. For the potential energy we note that for n stars V consists of $\frac{1}{2}n(n-1)$ interactions between pairs of stars. If in this crude picture we let each star have mass m, then a pair of stars at separation r contributes a potential energy $-Gm^2/r$, and for $n \gg 1$ the total potential energy of the cluster is

(3)
$$V = -n^2Gm^2/2R$$
$$= -GM^2/2R,$$

where R is the harmonic mean value of r over all pairs of stars. The quantity R is a sort of radius for the cluster; as a harmonic mean it clearly refers to the central bulk of the cluster rather than the outlying extremities.

Combination of Equations (1)—(3) gives

(4) $\langle v^2\rangle = GM/2R.$

Thus, incidentally, observations of the internal velocity dispersion of a stellar system, along with measurement of its radius, allow its mass to be determined.

The escape velocity can similarly be expressed in simple terms. For any given star the escape velocity is given by

(5) $v_e^2 = -2U,$

where U is the potential that applies to that star. The value of U is found from a summation over the other stars; a star of mass m at distance r contributes to this sum the quantity $-Gm/r$. Thus for n stars (we again ignore the difference between $n-1$ and n) there will be n such terms, and for an average over the cluster we can write

(6) $\langle v_e^2\rangle = 2GM/R,$

where R is again a harmonic mean distance between stars. Comparison with Equation (4) then gives

(7) $\langle v_e^2\rangle = 4\langle v^2\rangle.$

Although Equation (7) has been derived here in a crude schematic way, it can be shown to be rigorously true.

We can now define the characteristic time of mixing as the time that a typical star takes to move one cluster diameter. Thus the mixing time is

(8)
$$T_m = 2R/\langle v^2 \rangle^{1/2}$$
$$= (8R^3/GM)^{1/2}.$$

If we introduce the mean density,

(9)
$$\rho = 3M/4\pi R^3,$$

Equation (8) becomes

(10)
$$T_m = (6/\pi G\rho)^{1/2}.$$

The characteristic time of relaxation can be defined as the time required for the expected root-mean-square change in a star's kinetic energy to become as large as the average kinetic energy of a star. Calculation of the relaxation time involves first finding the energy exchange in a single encounter and then averaging over random values of five quantities that determine the circumstances of the encounter. The result is ([1, p. 67]; see also Woltjer's article, these Lectures)

(11)
$$T_r = 3^{1/2} \langle v^2 \rangle^{3/2}/16\pi^{1/2}G^2m^2N \ln(D_0\langle v^2 \rangle/2Gm),$$

where N is the number of stars per unit volume and D_0 is a cutoff distance beyond which encounters are no longer effective. Although D_0 was once taken to be the mean distance between stars, it is now recognized that distant encounters are fully effective even when simultaneous, so that all stars must be included. Thus we set $D_0 = 2R$; since it appears in the argument of a large logarithm, its choice need not be exact. Equation (4) now gives

(12)
$$D_0\langle v^2 \rangle/2Gm = n/2.$$

If we note that n can be written either as M/m or as $4\pi R^3N/3$, then Equation (11) becomes

(13)
$$T_r = (\pi nR^3)^{1/2}/(384\,Gm)^{1/2}\ln(n/2),$$

or again, in terms of density,

(14)
$$T_r = n/(512G\rho)^{1/2}\ln(n/2).$$

Thus slow relaxation goes with richness and especially with a large radius.

With these formulas, the ratio of relaxation time to mixing time becomes particularly simple:

(15) $$T_r/T_m = (\pi/3)^{1/2} n/32 \ln(n/2).$$

In a cluster with enough stars to be treated statistically, T_r is much greater than T_m; hence we may expect the relaxation effects to act slowly on stellar motions that remain well mixed at all times.

Before proceeding to the detailed dynamics of clusters we should take note of one more dynamical factor. The star clusters that we study in the Milky Way are all acted upon by the powerful force field of the Milky Way itself. This force as a whole merely affects the motion of the whole cluster and determines its orbit around the Galactic Center, but the differential force across the cluster acts as a stretching force. Although the galactic tidal force has very little effect on the dense center of the cluster, the outer parts are somewhat distorted. Most important, the galactic tidal force sets an upper limit on the distance at which the cluster can hold onto a star at all. The tidal effects are best examined in the framework of the restricted three-body problem [5, Chapter VIII]. In a coordinate system that rotates with the cluster's galactocentric motion, Jacobi's integral defines equipotential surfaces around the cluster center. Beyond the so-called straight-line Lagrangian points the surfaces open up, and stars with corresponding energies can escape. The distance of the Lagrangian points from the cluster center is, in galactic terms,

(16) $$r_t = R_g(M/3M_g)^{1/3},$$

where M_g is the mass of the galaxy and R_g is the distance of the cluster from the Galactic Center. The location of the tidal limit can thus be calculated, and in globular clusters the limit can be confirmed by studying counts of stars at different distances from the center.

Two physical processes thus determine radii for the cluster in two independent ways. First, the galactic tidal force sets a limiting radius, as just shown. Second, the negative binding energy of the cluster sets another radius, which is given by Equation (3). Since the dense central regions dominate the definition of this radius, it can be called the core radius. Clusters also differ in number of stars; thus it is clear that at least three parameters will be necessary to fix the density distribution in any given cluster. It is therefore quite significant that an extensive program of star counting shows [2] that three parameters are also *sufficient*—that is, that a single three-parameter formula fits the density distributions of all the clusters observed. Globular clusters are thus as similar as they can be, showing no noticeable differences that can be attributed to age or initial conditions. In other words, it appears that their forms are completely determined by the relaxation process.

Now for the detailed dynamics of star clusters. As we have seen, mixing is much more rapidly effective than relaxation, so that we may expect a cluster to be fully mixed at all times. That is, the cluster will be very close to a steady state, in which the velocity distribution and the density distribution are mutually consistent and do not change with time. As in Dr. Woltjer's paper in these Lectures, such a state is described by Liouville's equation, which is a continuity equation for the six-dimensional phase space of positions and velocities. The general stationary solution of this linear first-order partial differential equation states that the distribution function must be expressible as a function of the time-independent isolating integrals of the equations of motion of a star.

In this specific problem we assume spherical symmetry; hence spatially the distribution function can depend only on the distance from the center, r. The velocity distribution at any point, however, can depend on two variables, since the radial direction is uniquely distinguishable and the distribution function need only have axial symmetry about this direction. In other words, the radial and tangential components of velocity need not have the same distribution; and the distribution function of positions and velocities— the phase distribution—will thus depend on three variables. It turns out to be more convenient to replace the radial component of velocity by the magnitude of the velocity, v, so we will write the distribution function as $f(r, v, v_t)$, where v_t is the tangential velocity component.

In the present case of spherical symmetry, the integrals of the equations of motion of a star are

(17) $$E = v^2/2 + U(r),$$

(18) $$h = rv_t,$$

where E and h are constants that give the energy per unit mass and the angular momentum per unit mass, respectively, and U is the gravitational potential function. The general stationary solution of Liouville's equation then is given by

(19) $$f(r, v, v_t) = F(E, h),$$

where F is an arbitrary function. Although some restrictions are needed in order to avoid physical absurdities, a multiple infinity of functions F will lead to self-consistent cluster models. One can start with almost any density distribution and find a velocity distribution that fits, or alternatively one can choose a velocity distribution and find the corresponding density distribution.

Consideration of the mixing equilibrium thus leads, taken alone, to a hopeless indeterminacy. The key to our dilemma is the relaxation process, which leads to the choice of a particular velocity distribution. First, however, let us see how to go from a velocity distribution to a cluster model. Via Equations (17)—(19), the velocity distribution at a given point allows us to determine $F(E, h)$. The density at any point, ρ, can then be found by integrating over the velocity distribution. In a formal way this integration can be expressed as

(20) $$\rho = \int f d^2 v,$$

where the indicated integration is over the two velocity components on which f depends. This can now be transformed:

(21)
$$\rho = \int F d^2 v$$
$$= \int F \frac{\partial (v, v_t)}{\partial (E, h)} \, dE dh,$$

where the factor introduced is the Jacobian of the transformation expressed by Equations (17) and (18). The Jacobian introduces into the integrand the quantities r and $U - U_0$, where U_0 is the value of U at the point at which the velocity distribution was chosen. The result of the integration, carried out for all values of r and $U - U_0$, is a function $\rho(U - U_0, r)$. This function does not yet constitute a solution of the problem, however, since the dependence of U on r is still unknown. This dependence is found by solving Poisson's equation,

$$(22) \qquad \frac{d^2 U}{dr^2} + \frac{2}{r} \frac{dU}{dr} = 4\pi G \rho(U - U_0, r),$$

after which ρ can be expressed as a function of r alone.

This process becomes much simpler if the velocity distribution is everywhere isotropic—that is if it depends on v alone. Equation (19) is then replaced by

$$(23) \qquad f(r, v) = F(E).$$

Equations (20) and (21) involve only a single integration, and the density becomes $\rho(U - U_0)$. Poisson's equation must still be solved for $U(r)$, however.

To study the effect of relaxation on a velocity distribution, we now go to the other extreme—having considered mixing without relaxation, we now consider relaxation without mixing. Imagine a force-free region, in which the potential and the density are uniform. The effect of relaxation in such a region will be to make the velocity Maxwellian distribution. But as we have seen, the velocity distribution in a cluster cannot be Maxwellian because of the existence of an escape velocity. Hence we take our idealized relaxing region to be a square potential well, whose depth corresponds to an escape velocity beyond which stars leave irrevocably. The velocity distribution is now unable to become completely Maxwellian, even though it continually strives in that direction. Instead it settles down to a steady state in which the velocity distribution is nearly Maxwellian but drops to zero at the escape velocity. Stars are continually lost from the distribution at that end, while the distribution shifts in number without changing its shape.

The gradual changes induced by relaxation in a velocity distribution are described mathematically by the Fokker-Planck equation [7], which describes a diffusion in velocity space. For the steady-state problem the variables can be separated and the velocity equation solved numerically for the steady-state velocity distribution [8], whose exact form depends on where we set the escape-velocity cutoff.

On the one hand, then, mixing equilibrium permits almost any velocity distribution, which in turn determines the density distribution. On the other hand, our simplified example shows what kind

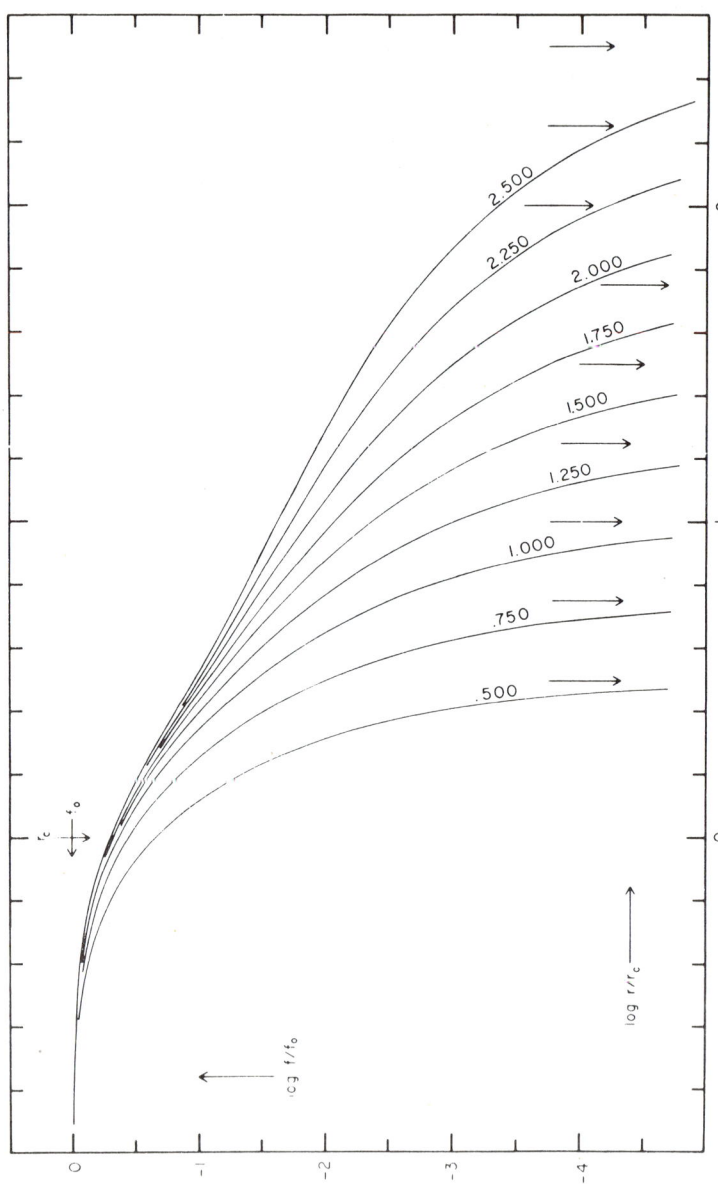

FIGURE 1. Surface density versus radius in theoretical cluster models.
Each curve is labeled with its value of $\log r_t/r_c$, and arrows mark limiting radii.

of velocity distribution is to be expected. The next step is to put the two together. As a first attempt we take the velocity distribution at the center of a cluster to be the steady-state solution found for the Fokker-Planck equation in a square-well potential. This fixes $F(E, 0)$, since at $r = 0$ all angular momenta are zero. The distribution function must still be chosen for other values of h; as the simplest first step we let F be independent of h, so that the velocity distribution is everywhere isotropic. The calculation of the cluster model then proceeds according to the simpler scheme following Equation (23).

The Fokker-Planck solutions constitute a one-parameter family of functions, according to where we set the escape cutoff—specifically, according to the value of $v_e^2 / \langle v^2 \rangle$ chosen for the cluster center; and the result is a one-parameter family of cluster models [3]. In each model the density falls to zero at a finite value of r, which becomes the limiting radius of the cluster. As $v_e^2 \to \infty$, so also does the limiting radius increase without limit, and the distributions of velocity and density approach the well-known correspondence between the Maxwellian velocity distribution and the isothermal sphere [1, p. 231].

The resulting density curves are shown in Figure 1. The quantity plotted is the surface density, found by parallel projection of the spatial density onto a plane—since this is the quantity that we actually observe in a star cluster. The individual curves are labeled by the distinguishing parameter $\log r_t / r_c$, where the core radius r_c is a scale factor that normalizes all the models to similar central density curves.

Although this calculation was intended as a crude first attempt at building cluster models, it succeeds surprisingly well, both in its agreement with observation and in the dynamical characteristics of the models. In the comparison with observation, first note that these models allow each cluster the requisite number of parameters. Since a scale factor and a number factor can clearly be chosen, a model can have a chosen limiting radius, core radius, and number of stars, just like a real cluster. In actual comparison with star counts in globular clusters, the agreement is quite satisfactory. Figures 2 and 3 show typical results in high- and low-concentration clusters, respectively. The vertical bars indicate the unavoidable statistical uncertainty in the counts. In Figure 2 the central region is missing because the crowding of stars made counting impossible, but the counts can be extended inward on less crowded photographs of shorter exposure. They show good agreement with the theoretical curve.

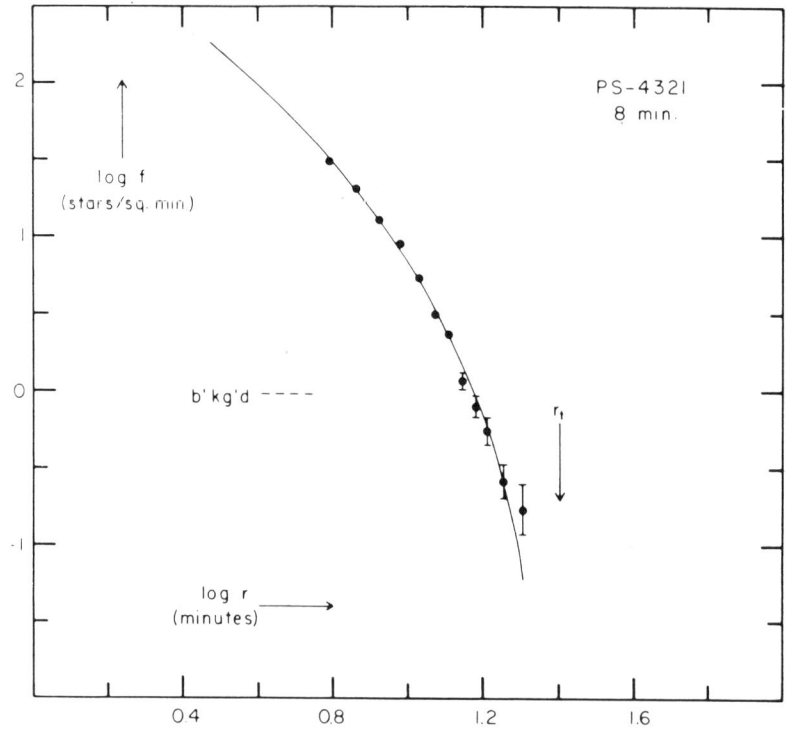

FIGURE 2. Star counts in the globular cluster M 13, compared with theoretical curve for $\log r_t/r_c = 1.50$. Maximum exposure with 48-inch Schmidt telescope of Palomar Observatory.

On the theoretical side we may examine the models from the point of view of relaxation theory. At the cluster center we have chosen the velocity distribution to fit the demands of relaxation, but how does it behave elsewhere? The answer to this question provides two happy surprises. First, the velocity distribution at *every* point in the cluster is very close to the steady-state solution of the Fokker-Planck equation corresponding to the escape velocity **at that point. Second, the corresponding loss rate of stars from**

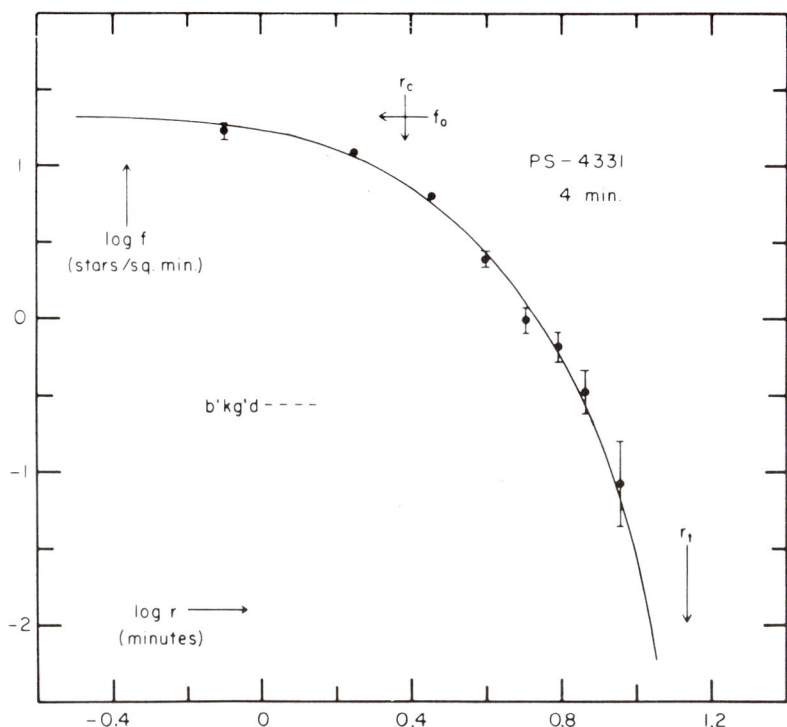

FIGURE 3. Star counts in the globular cluster NGC 5053,
compared with theoretical curve for $\log r_t/r_c = 0.75$.
Medium exposure with 48-inch Schmidt.

each point of the cluster is proportional to the density at that
point, so that the loss of stars does not demand an immediate
change in the equilibrium. (A change will in fact occur, as the
cluster adjusts to the overall decrease of star number; but such
a change will work on an evolution time scale rather than a relaxation time scale and is hence negligible in our present context.)

Application of relaxation theory thus leads to theoretical star-
cluster models that look as a cluster does and behave as a cluster
should. This is only a first step, however, and many other directions
need to be explored. For instance, models can be built on anisotropic
velocity distributions [4], and the effect of mixing stellar types
must be considered [6]. The basic dynamical processes in a star
cluster are understood, but we are far from a full knowledge of
these intriguing objects.

References

1. S. Chandrasekhar, *Principles of stellar dynamics,* Dover, New York, 1960.
2. I. R. King, Astronom. J. **67** (1962), 471.
3. ———, Astronom. J. **71** (1966), 64.
4. R. W. Michie, Monthly Notices Roy. Astronom. Soc. **125** (1963), 127.
5. F. R. Moulton, *An introduction to celestial mechanics,* Macmillan, New York, 1914.
6. J. H. Oort and G. van Herk, Bull. Astronom. Inst. Neth. **14** (1959), 229.
7. M. N. Rosenbluth, W. M. MacDonald and D. L. Judd, Phys. Rev. **107** (1957), 1.
8. L. Spitzer and R. Härm, Astrophys. J. **127** (1958), 544.

UNIVERSITY OF CALIFORNIA
BERKELEY, CALIFORNIA

D. Lynden-Bell

Cooperative Phenomena
in Stellar Dynamics

A cooperative phenomenon occurs whenever the behavior of a crowd of individuals is markedly different from a superposition of their individual behaviors.

Examples are: Plasma oscillations, two-stream instability, superconductivity, freezing and football crowds. We shall study such phenomena when our medium is made up of stars and interstellar gas. Because the field stated that way is far too large we shall limit ourselves to gravitational and magnetohydrodynamic phenomena only; it is quite likely that still more interesting phenomena will be discovered when the radiation of the stars coupled to thermal instabilities in the gas is considered. Thermal instability has correctly been emphasized as an important astronomical phenomenon in G. B. Field's recent comprehensive paper on the subject.

Stellar dynamics is, apart from a sign, very close to plasma physics, and most of the techniques I shall describe are borrowed from that subject. A short list of fundamental references is [1]—[13].

Except for [8] which is a development of [6] and [7] these are in order of publication. [9]—[12] concern themselves with oversimplified, infinite uniform media for which stability problems can be solved exactly. [6] − [8] concern themselves with the much harder real situation where the medium varies in density etc. from point to point—as all real systems do. I shall take these sections in reverse order, studying the idealised phenomena of the infinite medium first, and only then trying the inhomogeneous real situation.

The phenomena that we are led to discuss are Jeans's instability [9], [10], [11], Landau damping [9], two-stream instability [11], [13] and star-gas resonances [11], [13], while for the inhomogeneous systems we are interested in showing that sufficient inhomogeneity prevents Jeans's instability from occurring [6], [7], [8].

The method that we adopt follows the development of Wilson and Lynden-Bell.

I. Homogeneous Media

I.1. **Basic equations.** The Boltzmann-Liouville equation in rotating axes can be written

$$(1) \qquad \frac{\partial f_T}{\partial t} + \mathbf{c} \cdot \frac{\partial f_T}{\partial \mathbf{r}} + \mathbf{K}_T \cdot \frac{\partial f_T}{\partial \mathbf{c}} = 0.$$

$f_T d^3 r d^3 c$ is the total mass of stars in the phase space box of volume $d^3 r d^3 c$ about the point $\mathbf{r}, \mathbf{c}$ at time t, and

$$\mathbf{r} = (x, y, z), \qquad \mathbf{c} = (u, v, w),$$

$$\partial/\partial\mathbf{r} - (\partial/\partial x, \partial/\partial y, \partial/\partial z), \qquad \partial/\partial\mathbf{c} = (\partial/\partial u, \partial/\partial v, \partial/\partial w).$$

$\mathbf{K}_T$ is the total acceleration field,

$$(2) \qquad \mathbf{K}_T = \partial\psi_T/\partial\mathbf{r} - 2\mathbf{\Omega} \times \mathbf{c} + \Omega^2\mathbf{R},$$

ψ_T is the total gravitational potential,

$$\mathbf{\Omega} = (0, 0, \Omega), \qquad \mathbf{R} = (x, y, 0).$$

The gravitational potential is related to the density distribution by Poisson's equation

$$(3) \qquad \nabla^2\psi_T = -4\pi G(\rho_{T\text{stars}} + \rho_{T\text{gas}}),$$

where $\rho_{T\text{stars}} = \int f_T d^3 c$. Equations (1), (2), (3) determine the dynamics completely when there is no gas. When there is gas they must be supplemented by the equations of gas motion and continuity. Note that the individual masses of the stars do not appear because we have replaced the stars by a six-dimensional fluid in phase space and have ignored its graininess.

In these lectures we are concerned with the growth or decay of small perturbations from an equilibrium state. Let this state have potential Ψ, distribution function F and acceleration field $\mathbf{K}$ in rotating axes,

(4) $$\mathbf{K} = \partial \Psi / \partial \mathbf{r} - 2\Omega \times \mathbf{c} + \Omega^2 \mathbf{R}.$$

The equilibrium B-L equation reads

(5) $$\mathbf{c} \cdot \partial F / \partial \mathbf{r} + \mathbf{K} \cdot \partial F / \partial \mathbf{c} = 0.$$

We subtract the equilibrium equation from the total Equation (1) and define the perturbed quantities $\psi = \psi_T - \Psi$, $f = f_T - F$. We obtain, on linearisation in the perturbed quantities,

(6) $$\frac{\partial f}{\partial t} + \mathbf{c} \cdot \frac{\partial f}{\partial \mathbf{r}} + \frac{\partial \psi}{\partial \mathbf{r}} \cdot \frac{\partial F}{\partial \mathbf{c}} + \mathbf{K} \cdot \frac{\partial f}{\partial \mathbf{c}} = 0.$$

Similarly, subtracting the equilibrium Poisson equation from the total one,

(7) $$\nabla^2 \psi = -4\pi G \left(\int f d^3 c + \rho_{\text{gas}} \right),$$

where ρ_{gas} is the perturbation in the gas density. Equations (6) and (7) are the conventional starting point for stability analysis. We shall find an alternative set more appropriate for inhomogeneous systems in §II, but along the present line we can go no further without specialising to some definite equilibrium state, so that F and $\mathbf{K}$ take specific forms.

I.2. **Homogeneous equilibria.** In a homogeneous equilibrium the acceleration field $\mathbf{K}$ must be independent of position. Now

(8) $$\mathbf{K} = \partial \Psi / \partial \mathbf{r} - 2\Omega \times \mathbf{c} + \Omega^2 \mathbf{R},$$

which can be made independent of position if the two space dependent terms cancel, i.e.,

(9) $$\partial \Psi / \partial \mathbf{r} = -\Omega^2 \mathbf{R}, \qquad \mathbf{K} = -2\Omega \times \mathbf{c},$$

which yield on taking the divergence

(10) $$\nabla^2 \Psi = -2\Omega^2.$$

Comparison with the unperturbed Poisson equation shows that

the total equilibrium density must be $\Omega^2/2\pi G$. With this choice of $\mathbf{K}$, can we find a solution for the equilibrium distribution function which is homogeneous? Any homogeneous F satisfying Equation (5) with our special form for $\mathbf{K}$ must satisfy

$$(-2\Omega \times \mathbf{c}) \cdot \partial F/\partial \mathbf{c} = 0.$$

If we take cylindrical polar coordinates in velocity space by writing

(11) $$\mathbf{c} = (c_\perp \cos\phi_c, c_\perp \cos\phi_c, w),$$

then

$$(\Omega \times \mathbf{c}) \cdot \partial/\partial \mathbf{c} = \Omega\, \partial/\partial\phi_c,$$

so

$$\partial F/\partial\phi_c = 0,$$

and consequently

$$F = F(c_\perp, w).$$

These equilibrium distribution functions give us a uniform density, and if there is no other material present supplying the gravity field, we must have

$$2\pi \iint F(c_\perp, w)\, c_\perp\, dc_\perp\, dw = \frac{\Omega^2}{2\pi G},$$

the equilibrium density. However, in general there will be other material present, so we shall not use this relation which can be thought of as a normalisation condition on F.

I.3. Fourier-Laplace transformation. Equation (6) now depends on position only through the perturbed quantities, in which it is linear. Hence on Fourier transformation each coefficient separates off, giving us equations for the coefficient of $\exp(i\mathbf{k} \cdot \mathbf{r})$ of the form

(12) $$\frac{\partial f_\mathbf{k}}{\partial t} + i\mathbf{k} \cdot \mathbf{c} f_\mathbf{k} + i\mathbf{k} \cdot \frac{\partial F}{\partial \mathbf{c}}\, \psi_\mathbf{k} - 2\Omega\frac{\partial f_\mathbf{k}}{\partial\phi_c} = 0,$$

(13) $$\frac{k^2}{4\pi G}\psi_\mathbf{k} = \int f_\mathbf{k}\, d^3c + \rho_{g\mathbf{k}}.$$

Suffixes $\mathbf{k}$ denote the kth Fourier component. Since no direction in the x, y plane is preferred we may orient the y-axis along the component of $\mathbf{k}$ in that plane. At the same time we find the Laplace transform in time by multiplying by e^{-st} and integrating with respect to t. We take $\text{Re}(s)$ so large that the integrals converge and denote

$$\int_0^\infty f_{\mathbf{k}} e^{-st} dt \quad \text{by} \quad \tilde{f}(\mathbf{k}, \mathbf{c}, s) \quad \text{etc.}$$

Our equations then become, for $\text{Re}(s)$ large:

(14) $$(s + ik_z w + ik_\perp c_\perp \sin \phi_c) \tilde{f} - 2\Omega \frac{\partial \tilde{f}}{\partial \phi_c} = g - i\mathbf{k} \cdot \frac{\partial F}{\partial \mathbf{c}} \tilde{\psi},$$

(15) $$\frac{k^2}{4\pi G} \tilde{\psi} = \int \tilde{f} d^3 c + \tilde{\rho}_g,$$

where the first term has been integrated by parts and $g(\mathbf{k}, \mathbf{c})$ is the initial value of $f_{\mathbf{k}}$ at time $t = 0$.

To understand any complicated problem it is usually easiest to solve all the neighboring oversimplified problems first. One of these is obtained by putting $\Omega = 0.$[1] However, one may show that this situation gives the same equations as we get by treating waves with $\mathbf{k}$ along Ω.

Motions perpendicular to Ω continue their unperturbed way, so we shall integrate over these.

I.4. **Phenomena occurring without rotation.** We take $\mathbf{k} = (0, 0, k)$ and integrate Equation (14) with respect to ϕ_c and $c_\perp$; we also define

$$\tilde{\bar{f}} = \int \int \tilde{f} d\phi_c c_\perp \, dc_\perp$$

and $\bar{g}$, $\bar{F}$ similarly and remember that all physical quantities are periodic in ϕ_c.

Equation (14) then reads

(16) $$(s + ikw) \tilde{\bar{f}} = g - ik (\partial \bar{F}/\partial w) \tilde{\psi}.$$

The w-dependence of the right-hand side is known even though $\tilde{\psi}$ is unknown. We may find the w-dependence of $\tilde{\bar{f}}$ by division:

(17) $$\tilde{\bar{f}} = \frac{\bar{g} - ik (\partial \bar{F}/\partial w) \tilde{\psi}}{s + ikw}.$$

---

[1] There is then no real equilibrium since the density must vanish.

To go further we must relate the potential to its sources by using Poisson's Equation (15). For the present we put $\rho_g = 0$, and it becomes

$$\frac{k^2}{4\pi G} \tilde{\psi} = \int \frac{\overline{g}}{s + ikw} \, dw - \int \frac{ik}{s + ikw} \frac{\partial \overline{F}}{\partial w} \, dw \, \tilde{\psi}.$$

Thus

(18)
$$\tilde{\psi} = \frac{I_g}{k^2/4\pi G + I}$$

where I_g and I are the integrals of the line above. Formally Equation (18) is the solution to our problem since the right-hand side is known. Furthermore, substituting this $\tilde{\psi}$ back into Equation (17) we obtain the solution for the distribution function. However, to obtain formally the Fourier-Laplace transform of the desired solution is one thing, and to understand its complications is another.

I.5. **Determination of stability by Laplace transformation.** We wish to invert our Laplace transformations to obtain the temporal behavior of $\psi_k(t)$; this is done by the inversion formula

$$\psi_k(t) = \frac{1}{2\pi i} \int_{\sigma - i\infty}^{\sigma + i\infty} \tilde{\psi}(s, \mathbf{k}) \, e^{st} ds,$$

where σ is to be taken large and positive so that we only require $\tilde{\psi}$ for values of s with $\mathrm{Re}(s)$ large, This is the condition under which we derived our expression for $\tilde{\psi}$. However, it is very instructive to analytically continue $\tilde{\psi}$ to values of s with $\mathrm{Re}(s)$ no longer large. At some values of the complex variables $\hat{\sigma}(s)$ will have poles; we denote these by $\odot$. The original path of integration along the solid path C may be transformed by Cauchy's theorem to the dotted line path. The pieces at infinity vanish because e^{st} fluctuates violently, so we are left with the integral up the straight dotted line C_1 plus the contributions from the poles, thus

$$\psi_k(t) = \frac{1}{2\pi i} \int_{C_1} \tilde{\psi} e^{st} ds + \sum_r R_r e^{s_r t},$$

where R_r is the residue at the pole S_r of the function $\tilde{\psi}$. From this form it is obvious that for large t the expression is dominated by the contribution from the pole furthest to the right in the complex

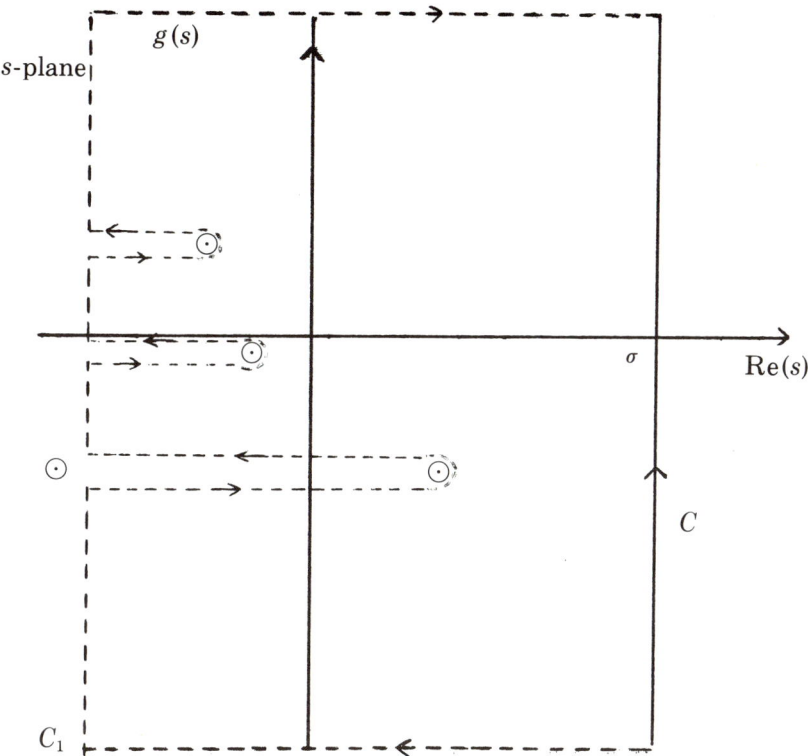

FIGURE 1. Singularities of $\tilde{\psi}$ and deformation of integration path, $C \rightarrow C_1$

s-plane; the amplitude grows like $R^{\text{Re}(s)t}$. In particular, if there are no poles of ψ with $\text{Re}(s) > 0$, then the perturbation does not grow, while if there are such poles the system is unstable. Thus the necessary and sufficient condition for instability is that there should be a pole (or at least a singularity) of $\tilde{\psi}$ at a point with $\text{Re}(s) > 0$.

We now aim to look at the analytic behavior of $\tilde{\psi}$, so that we can continue it for $\text{Re}(s)$ not large and determine whether any such singularities exist. We shall assume that $\overline{g}(w)$ and $\partial \overline{F}(w)/\partial w$ are analytic functions of w. Then

$$I_g(\zeta) = \frac{1}{ik} \int \frac{\overline{g}(w)}{w - \zeta} \, dw, \qquad \zeta = \frac{is}{k},$$

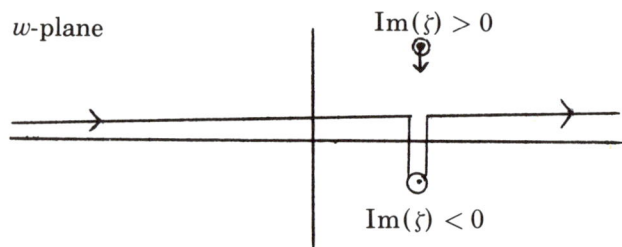

FIGURE 2

is analytically defined everywhere provided we take the integral from $-\infty$ to ∞ beneath the pole $w = \zeta$ (note that it is beneath for $\mathrm{Re}(s)$ large; if we define it to be beneath always, then we have the analytic continuation, see Figure 2). The same is true of the other integral

$$(19) \qquad I(\zeta) = \int_{\text{under pole}} \frac{\partial \overline{F}(w)/\partial w}{w - \zeta}\, dw, \qquad \zeta = \frac{is}{k};$$

thus for $\mathrm{Re}(s)$ (or $\mathrm{Im}(\zeta)$) > 0

$$I(\zeta) = \int_{-\infty}^{+\infty} \frac{\partial \overline{F}(w)/\partial w}{w - \zeta}\, dw,$$

and for $\mathrm{Re}(s) < 0$ we must add to this expression $2\pi i (\partial \overline{F}(w)/\partial w)_{w=\zeta}$ which is the contribution from the pole (see Figure 2). As ζ decreases through the real axis it has to drag the contour down, if the resulting integral is to be analytically continued. With these analytic continuations of I and I_g we see that neither have poles, hence the only possible poles of $\tilde{\psi}$ are the zeros of

$$k^2/4\pi G + I.$$

Instability only occurs when there are such zeros in the region

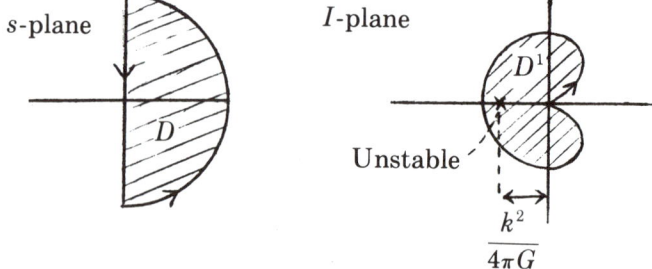

FIGURE 3. The mapping $s \to I$

$\mathrm{Re}(s) > 0$. The boundary of this region, $\mathrm{Re}(s) = 0$, is the line on which ς is real. The function $I = I(is/k)$ defines a mapping of the complex s-plane into the complex I-plane. As one passes around the domain D in the complex s-plane which includes all of $\mathrm{Re}(s) > 0$, one performs some contour in the I-plane which includes a domain D^1 on its left-hand side. This domain is the domain of those values of I attained when s is in D. Thus there are zeros of $k^2/4\pi G + I$ in D if and only if D^1 contains the point $-k^2/4\pi G$. So we have reduced our problem of stability to tracing the curve $I(\varsigma)$ when ς is real.

I.6. **Jeans's instability.** For real ς we can write

$$
\begin{aligned}
I(\varsigma) &= \int_{-\infty}^{\infty} \frac{\partial \overline{F}/\partial w}{w - \varsigma}\, dw \qquad \text{under } w = \varsigma \\
&= \int_{-\infty}^{\infty} \frac{\partial \overline{F}/\partial w}{w - \varsigma}\, dw + i\pi \left(\frac{\partial \overline{F}}{\partial w}\right)_{\varsigma} \\
&= \int_{-\infty}^{+\infty} \frac{\overline{F}(w) - \overline{F}(\varsigma)}{(w - \varsigma)^2}\, dw + i\pi \left(\frac{\partial \overline{F}}{\partial w}\right)_{\varsigma}.
\end{aligned}
\tag{20}
$$

The crossed integral denotes a Cauchy principal value. $I(\varsigma)$ can only be real with ς real when $(\partial \overline{F}/\partial w)_{\varsigma} = 0$, that is, when ς is the velocity of a turning point in the distribution function; evidently when $\overline{F}(\varsigma)$ is a maximum $I(\varsigma)$ is real and negative. Thus for functions $\overline{F}(w)$ with a single maximum at $w = \varsigma$ there will be instability at all wave numbers k smaller than those given by

$$
\frac{k^2}{4\pi G} = \int_{-\infty}^{\infty} \frac{\overline{F}(\varsigma) - \overline{F}(w)}{(w - \varsigma)^2}\, dw.
\tag{21}
$$

Notice that the integral is large if $F(w)$ has a large curvature at its maximum. For a Maxwellian at rest the integral reduces to $\rho_0/2c_s^2$, where ρ_0 is the unperturbed stellar density and $\overline{F}(w) = \rho_0(2\pi c_s^2)^{-1/2}\exp(-w^2/2c_s^2)$ which defines c_s^2. A Maxwellian distribution of stars is therefore unstable to waves of wave number k if

$$
k^2 c_s^2 < 4\pi G\rho,
\tag{22}
$$

which is precisely Jeans's criterion.

For a superposition of two Maxwellians at rest it is clear that $k^2/4\pi G$ is linear in $\overline{F}$, so the required condition is

$$
k^2/4\pi G < \rho_1/c_1^2 + \rho_2/c_2^2.
$$

If we define $\bar{\rho}$ and $\bar{c}$ by

(23)
$$\bar{\rho} = \rho_1 + \rho_2,$$
$$\bar{\rho}/\bar{c}^2 = \rho_1/c_1^2 + \rho_2/c_2^2,$$

we obtain $k^2\bar{c}^2 < 4\pi G\bar{\rho}$. This shows that the correct density for Jeans's criterion is the sum of the densities, and the correct c^2 is the harmonic, density weighted mean. This result is important, for it shows that small amounts of material with low velocity dispersion can be important in determining stability. In particular it is quite likely that the interstellar gas, which is but a small proportion of the matter density, may, through its low velocity dispersion, be almost as important as the stars as a cause of Jeans's instability.

I.7. **Landau damping.** Whereas in frictionless gas dynamics the stable waves are normal, undamped sound waves, a more interesting phenomenon occurs in plasmas and in stellar dynamics, where the stable waves damp by feeding their cooperative wave energy into individual particle energy. Since we have already done most of the mathematics, I will first derive the result and then discuss its physical significance.

For stable waves the most important pole is either on the left of the real axis or on it. When Jeans's criterion is not even marginally satisfied one may show that the pole is on the left, so

$$I = \int_{-\infty}^{\infty} \frac{\partial \bar{F}/\partial w}{w - \zeta}\,dw + 2\pi i \left(\frac{\partial \bar{F}}{\partial w}\right)_\zeta .$$

$\zeta = is/k = -\omega/k$, if we define ω to be the "frequency of the disturbance" (at present the contribution from our most important pole behaves like e^{st}). The condition that we have a pole,

(24)
$$I(-\omega/k) = -k^2/4\pi G,$$

now becomes the dispersion relation connecting the values of ω and k of the free waves that can travel through the medium. The fact that the pole is on the left is obvious since there are no solutions with real ω except the marginal one at $\omega = 0$. If we solve Equation (24) with large k and a Maxwellian $F(w)$, we find very approximately

$$\omega \approx kc_s(L + i\,\pi/4L),$$

where

$$(25) \qquad L^2 \approx \log \left| \frac{k^2 c_s^2}{4\pi G \rho_0} \cdot \frac{1}{2\pi^{1/2}} \right|.$$

Notice that the damping rate is a property of the unperturbed medium; it is independent of the relative phases of the different applied perturbations. The importance of these would be very apparent if we analysed I_g in detail.

I.8. **Physical discussion.** Let us first discuss what would happen to a density disturbance of wave number k if there were no inter-action at all, $G = 0$. Each particle would move, and in a time of $\sim 1/kc_s$ the fact that the disturbed particles had a velocity dis-persion of about c_s would considerably lessen the density disturbance. This mere mixing is an important agency in bringing stellar systems to look smooth; a nice example of the process is afforded by particles in a pig trough viewed end on:

Release the particles how you like, with some initial distribution function $f(E, \phi, 0)$, where E is the energy of the oscillation and ϕ its phase. Now plot the phase space, E against ϕ, and assume that—like in most dynamical systems—the larger amplitudes have the longer periods.

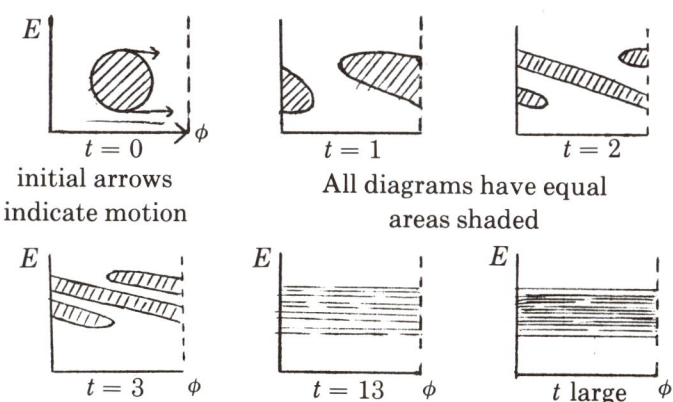

FIGURE 4. Evolution of a distribution function in phase space
due to mere propagation without interactions

A mathematical investigation shows that, for $t \to \infty$, $f(E, \phi, t)$ is not pointwise convergent—but it does converge in the mean to the average $\bar{f}(E)$ of $f(E, \phi, 0)$ over all phases, i.e., $\int f Q d\tau \to \int \bar{f} Q d\tau$

for smooth Q. But $G \neq 0$, the density disturbances associated with phase mixing cause gravity disturbances, and these are themselves damped by interactions with the medium. The detailed mechanism of Landau damping is that particles moving just a little more slowly than the wave gain energy from it on the average, while those moving more rapidly lose energy. A brief discussion is given in Stix [2]. It should be possible to give a nonlinear theory of Landau damping along these lines. In stellar dynamics Landau damping is probably the last phase of the most efficient relaxation process, which I call mean field relaxation; this only occurs when a stellar system is falling about, before it is in a steady state. The associated relaxation time for an individual to gain or lose energy is $(G\rho)^{-1/2}$— *exceedingly short*. This process is very important, not understood at all, and a good thing to work on. It is known that it does not lead to a Maxwellian distribution, but to something rather more like a Fermi-Dirac one.

I.9. **Relaxation or smoothing processes.** We just list the relevant processes:

(1) Phase mixing—mere propagation. The characteristic time is that for which individuals get out of step.

(2) Mean field relaxation. Time scale $(G\rho)^{-1/2}$. The last phases are probably the same as (3).

(3) Landau damping; gain of particle energy from waves just discussed.

(4) Normal relaxation due to the graininess of the individual particles that make up the medium. Time scale 10^{13} years or more for stars in the galaxy, but of interest in globular clusters.

I.10. **Two-stream instability.** If we have two groups of stars (that is stars associated in velocity, but not in position) and they interpenetrate, are there any new instabilities due to the relative velocities of the streams? We have

$$I_T = I_{\text{total}}(\zeta) = \int_{-\infty}^{+\infty} \frac{\partial \overline{F}_{\text{total}}/\partial w}{w - \zeta} dw$$

$$= \int_{-\infty}^{+\infty} \frac{\partial F_1/\partial w + \partial F_2/\partial w}{w - \zeta} dw$$

$$= I_1(\zeta) + I_2(\zeta).$$

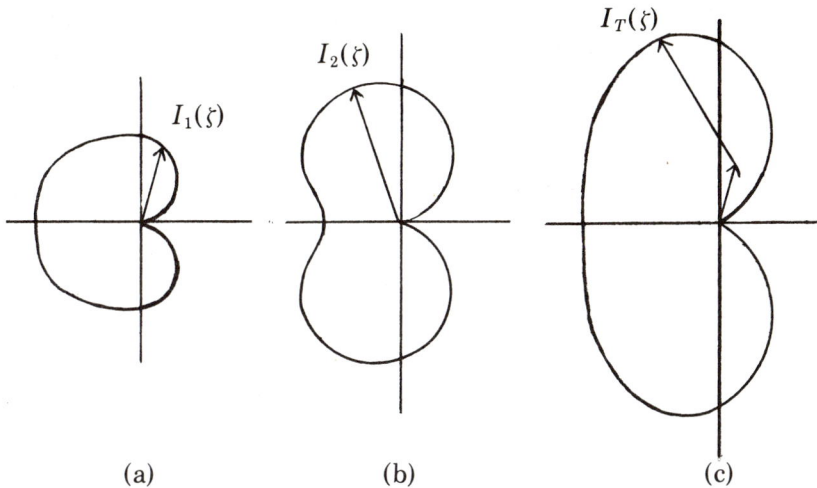

FIGURE 5. Superposition of the curves $I_1(\zeta)$, $I_2(\zeta)$, ζ real

Now suppose we move the distribution F_2 with velocity W. To do this we write $F_2(w - W)$ for $F_2(w)$, and we write I_2^1 for the new I_2:

$$I_2^1(\zeta) = \int_{-\infty}^{\infty} \frac{\partial F_2(w - W)/\partial w}{(w - \zeta)} \, dw$$

$$= \int_{-\infty}^{\infty} \frac{\partial F_2(w)/\partial w}{w - (\zeta + W)} \, dw = I_2(\zeta + W).$$

Thus the only effect of movement is to keep the function I_2 the same, but to relabel the points on the curve in the I-plane with new values of ζ.

To get the most unstable situation we need the largest possible k to be marginally stable.

Obviously

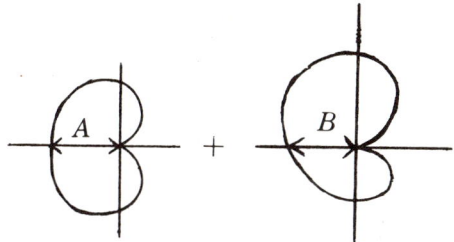

cannot get a larger value of $k^2/4\pi G$ than $A + B$. However, equally obviously,

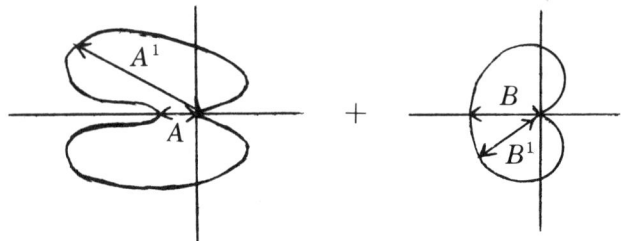

possibly gives a resultant

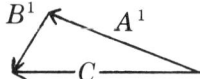

and we should be able to gain by making the two primed arrows correspond to the same ζ; for then, instead of $A + B$ we will have C which can be greater. We can choose which points correspond to a given ζ by suitably prescribing the mean velocities with which the star groups move. Thus there are definitely situations in which relative motion of star groups each of which is itself symmetrical, can cause a wave length to be unstable which would not be if the groups were at relative rest. However, this phenomenon cannot occur with two Maxwellian groups, since neither of these has the required concavity on the left. In fact, no distribution with $\partial^2 F/\partial (w^2)^2 > 0$ has such a concavity in its I-diagram. Such distributions are very flat topped. Only when one stream is very flat topped can the movement of the other cause the more peaky top required for two-stream instability (see our earlier remark that high curvature at the maximum promotes instability; Equation (21)). This is easiest understood diagramatically:

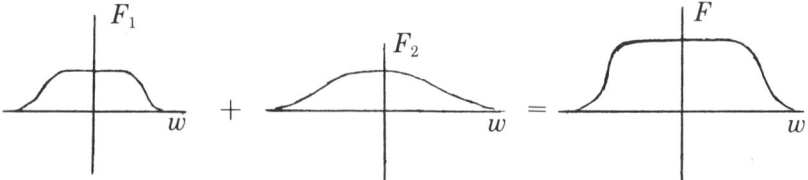

But with motion a more peaky top may be obtained; moving F_2 to the right gives

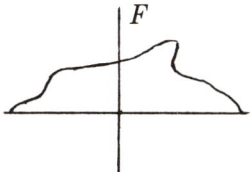

With two rather flat topped distributions the increase in peakiness with relative motion is even more pronounced as the reader can easily imagine.

I.11. **Interactions between gas and stars.** We again consider waves along Oz. The perturbed equations of continuity, state, and gas motion are:

$$\partial \rho_g / \partial t + \partial (\rho_{og} u) / \partial z = 0,$$

$$P_g = c_s^2 \rho_g,$$

$$\frac{\partial u}{\partial t} = -\frac{1}{\rho_{og}} \frac{\partial P_g}{\partial z} + \frac{\partial \psi}{\partial z}.$$

Eliminating P_g and Fourier-Laplace transforming we obtain

$$s\widetilde{\rho}_g + ik\rho_{og}\widetilde{u} = \rho_g(0),$$

$$s\widetilde{u} + c_s^2 ik\widetilde{\rho}_g / \rho_{og} - ik\widetilde{\psi} = u(0).$$

Eliminating $\widetilde{u}$ results in

$$(s^2 + k^2 c_s^2)\widetilde{\rho}_g = k^2 \rho_{og}\widetilde{\psi} + [s\rho_g(0) - ik\rho_{og}u(0)].$$

If we insert this value of $\widetilde{\rho}_g$ and expression (17) in the Poisson Equation (15), we find

$$\frac{k^2}{4\pi G}\widetilde{\psi} = I_g - I\widetilde{\psi} + \frac{k^2 \rho_{og}}{s^2 + k^2 c_s^2}\widetilde{\psi} + \frac{[\quad\quad]}{s^2 + k^2 c_s^2},$$

where the empty bracket contains only initial data; hence

$$\widetilde{\psi} = \frac{I_g + [\quad]/(s^2 + k^2 c_s^2)}{k^2/4\pi G + I - k^2 \rho_{og}/(s^2 + k^2 c_s^2)}.$$

Once again the numerator depends only on initial conditions, and there are no poles with $\mathrm{Re}(s) > 0$ in the s plane other than those arising from zeros of the denominator. We therefore study the denominator $k^2/4\pi G + \int ((\partial \overline{F}/\partial w)/(w - is/k))\,dw - \rho_{og}/(c_s^2 + s^2/k^2)$.

Now

$$\frac{\rho_{og}}{c_s^2 + s^2/k^2} = \frac{\rho_{og}}{2c_s}\left[\frac{1}{c_s - is/k} + \frac{1}{c_s + is/k}\right]$$

$$= \frac{\rho_{og}}{2c_s}\int\left(\frac{\delta(w - c_s)}{w - is/k} - \frac{\delta(w + c_s)}{w - is/k}\right)dw.$$

Now define the mesa function of area ρ_{og} by

$$M(w) = \rho_{og}/2c_s \quad \text{if } |w| \leq c_s$$

$$= 0 \quad\quad \text{if } |w| > c_s,$$

then

$$\frac{\rho_{og}}{c_s^2 + s^2/k^2} = \int\frac{-\partial M/\partial w}{w - is/k}dw.$$

Hence our denominator is

$$\frac{k^2}{4\pi G} + \int\frac{\partial(\overline{F} + M)/\partial w}{w - is/k}dw.$$

This demonstrates that the system of gas and stars behaves in precisely the same way as a system of stars alone whose distribution function is not F, but $F + M$. We may now use our analysis of the purely stellar case to discuss interactions with gas.

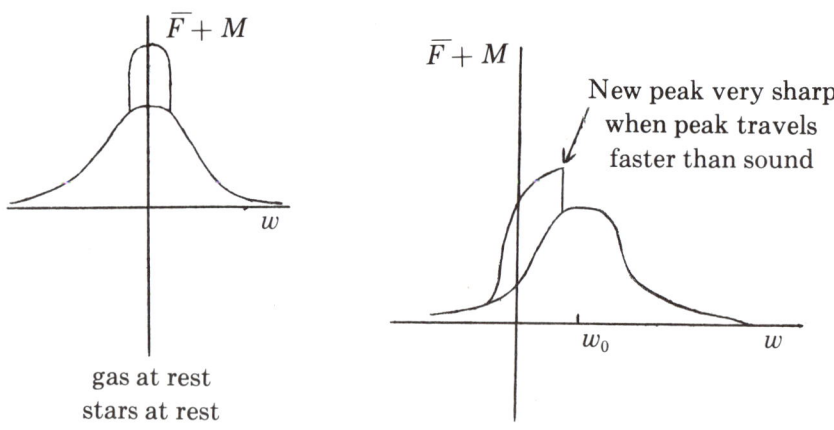

FIGURE 6. Stellar (pseudo-) distributions, modified by gas

In the Nyquist diagrams the situation looks like this:

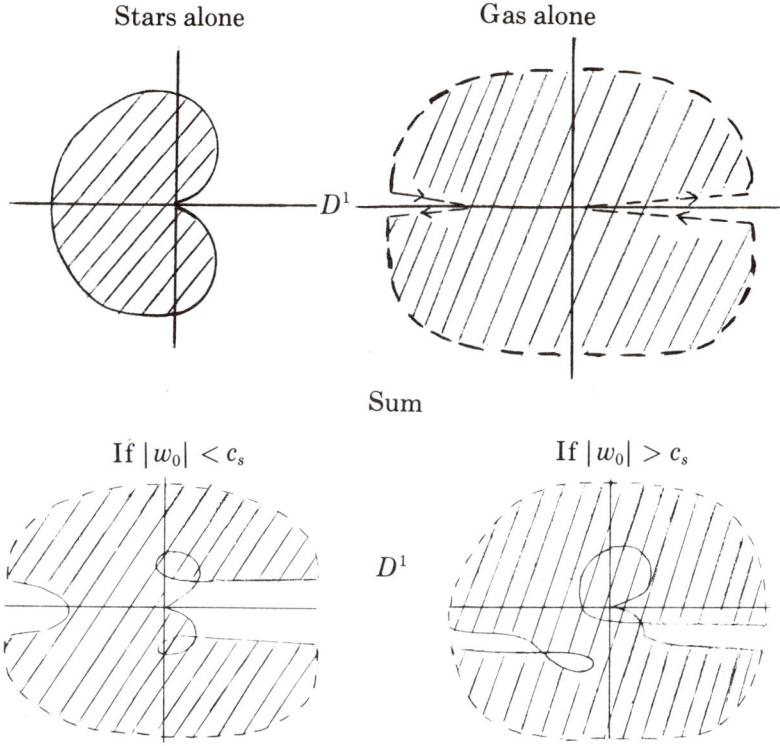

FIGURE 7. The regions D^1 (cf. Figure 3) in the I-plane for a stellar system interacting with gas; w_0 is defined by
$$\overline{F}(w_0) = \text{Max}_w \, \overline{F}(w).$$

In the case $|w_0| > c_s$ the whole negative real axis is contained in D^1. Thus we have instability at all wave lengths whenever the velocity w_0 of the peak of the stellar distribution function exceeds the sound velocity c_s of the gas, a result first obtained by Sweet [11].

I.12. **Physical interpretation and discussion of star-gas resonances.** These phenomena are easily understood in terms of stars riding the gravity field of the gaseous sound wave. Those going just faster give energy to the wave, those going just more slowly take energy from it. If therefore there are more stars going just faster than sound they form a cooperative sonic boom which is built up in the gas. This is the "music of the stars."

If the gas contains a magnetic field then there are two possible velocities of propagation at the modified sound-Alfvén speeds. Stars can resonate with each of these, and the appropriate effective distribution function for the magnetohydrodynamic gas is a double mesa

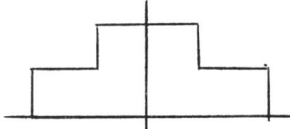

the cliffs coming at the different sound speeds.

As one might expect of a purely gravitational interaction all these instabilities grow only on the $(G\rho)^{-1/2}$ time scale—the actual growth rate is even lower since it is only that part of ρ that contributes to the extra instability beyond Jeans's that is effective. The growth times for the galaxy are $\sim 5 \cdot 10^8$ years or more, and it is not clear that in practice a star-sound wave resonance could be maintained for anything like that long. To complicate matters still farther it is unlikely that stars performing Lindblad orbits can keep in resonance with any sound wave for long enough that the growth is appreciable. Thus although cooperative instabilities are a mechanism which one might expect to make the star and the gas move together—at least to within the gas's velocity of sound, nevertheless it is very doubtful that this mechanism can actually be effective in coupling gas and stars.

I.13. **Rotation-dependent phenomena.** In this section we will not specialise to waves in any particular direction, but will give an indication of how to solve the general case represented by Equations (14) and (15). Equation (14) may be integrated by using the integrating factor

$$h = \exp\{(-1/2\Omega)\left[(s + ik_z w)\,\phi_c - ik_\perp c_\perp \cos\phi_c\right]\}.$$

When the boundary condition that everything must be periodic in ϕ_c is applied one obtains the solution

$$\tilde{f} = (2\Omega h)^{-1}\left\{\tilde{\psi}i\mathbf{k} \cdot \int_{-\infty}^{\phi_c} \frac{\partial F}{\partial \mathbf{c}}\, h\,d\phi_c - \int_{-\infty}^{\phi_c} gh\,d\phi_c\right\}.$$

Rather tediously one may now integrate over all velocities and put the result into Equation (15). Taking $\rho_g = 0$ one may then solve for $\tilde{\psi}$ and determine stability. However the integrations are difficult, and some form of distribution function $F(c_\perp, w)$ has to be assumed. So far it is known that isotropic distributions have only Jeans's instability, and no other instability has been discovered although instabilities analogous to Harris's[2] instabilities are likely to occur when the distribution functions have too great an anisotropy. W. B. Wilson is at present investigating such effects. For systems of stars and gas resonances occur not only when $c_s = \omega/k$ but also when $c_s = (\omega + 2n\Omega)/k$ where n is integral, but the process is much the same.

For an isotropic Maxwellian one can show that Jeans's instability occurs when $k^2 c_s^2 < 4\pi G\rho$, independently of the direction of $\mathbf{k}$ unless it is perpendicular to Ω when the criterion is modified to

$$\frac{k^2}{4\pi G\rho_0} = 1 - I_0\left(\frac{k_\perp c_\perp}{2\Omega}\right)^2 \exp\left\{-\left(\frac{k_\perp c_\perp}{2\Omega}\right)^2\right\}.$$

I.14. **Toomre's problem.** An ingeneous way of getting a homogeneous medium and a problem of stability of some importance is to take a model of the galaxy that is perfectly flat, then to take a small piece of it and to look for instabilities. We deal with two-dimensional distribution functions $F(c_\perp)$, and $\mathbf{k}$ is always perpendicular to Ω. The whole of our former mathematics may be used except that Poisson's equation must be modified to

$$\left.\frac{\partial\psi}{\partial z}\right|_{z=0-}^{z=0+} = 4\pi G\int f(c_\perp)\,d^2c_\perp$$

together with $\nabla^2\psi = 0$ if $z \neq 0$.

It follows that $\psi(\mathbf{R}, z)$ is a superposition of terms proportional to $e^{-|k||z|}e^{i\mathbf{k}\cdot\mathbf{R}}$. Let us put $z = 0$ in $\psi(\mathbf{R}, z, t)$ and Fourier-transform with respect to $\mathbf{R}$ ($\mathbf{R}\cdot\Omega = 0$) and Laplace-transform in t, obtaining $\tilde{\psi}(\mathbf{k}, s)$. Then accordingly the transformed Poisson equation reads

$$-|k|\,\tilde{\psi} = -2\pi G\int \tilde{f}(c_\perp)\,d^2c_\perp.$$

The net effect is that $|\mathbf{k}|/2\pi G$ replaces the $k^2/4\pi G$ of our former treatment and our distribution functions are two-dimensional, our densities surface densities. Jeans's criterion for a nonrotating

---

[2] In plasma physics this instability can occur when the dispersion in $c_\perp$ is more than twice that in w; it occurs for wave vectors neither parallel nor perpendicular to Ω.

flat sheet becomes

$$|\mathbf{k}|/2\pi G < \Sigma_0/c_s^2,$$

where Σ_0 is the unperturbed density. For two Maxwellians it is again

$$|\mathbf{k}|/2\pi G < \Sigma_1/c_1^2 + \Sigma_2/c_2^2,$$

again demonstrating the importance of material with low velocity dispersion.

However, for thin sheets rotation has a dominant stabilising influence at long wave lengths. For a gaseous sheet the dispersion relation is

$$\frac{|\mathbf{k}|}{2\pi G} = \frac{\Sigma_0}{c_s^2 + (4\Omega^2 - \omega^2)/k^2},$$

and the equation for the critical wave numbers is

$$\frac{k}{2\pi G} = \frac{\Sigma_0}{c_s^2 + 4\Omega^2/k^2}, \text{ i.e., } k^2 c_s^2 - 2\pi G\Sigma_0 k + 4\Omega^2 = 0$$

or

$$\left(k - \frac{\pi G\Sigma_0}{c_s^2}\right)^2 + \left(\frac{4\Omega^2}{c_s^2} - \frac{\pi^2 G^2\Sigma_0^2}{c_s^4}\right) = 0,$$

which shows that there are only unstable waves when $2\Omega < \pi G\Sigma_0/c_s$. For a stellar sheet Toomre obtained a result that only differs from this in that 3.36 replaces π.

In all the cases we have so far considered Jeans's instability is almost or exactly the same for stellar and for gaseous systems with the convention that a Maxwellian is written $\exp[-w^2/2c_s^2]$ and c_s is replaced by the sound velocity in the gaseous case. We shall see presently that there are reasons for believing this even in inhomogeneous systems.

II. INHOMOGENEOUS MEDIA AND THE SUPPRESSION OF JEANS'S INSTABILITY

A. **Stability of stellar systems.** An energy principle which gives a sufficient condition for the stability of a stellar system is derived. There is a similar principle in self-gravitating gas dynamics which is both necessary and sufficient. Comparison enables one to infer the stability of a stellar system from that of a gaseous system with the same density distribution.

A.1. *Introduction.* The first progress in the discussion of collisionless stability of realistic stellar systems was made when Antonov published his necessary and sufficient variational principle [6] and used it to demonstrate the stability of the finite "polytropic" systems [7]. In the West the discussion of the infinite uniform media [9] − [11], [14], [15] had hardly begun and was mainly the translation of plasma physical results into the stellar dynamical language. However such methods are not of great power for very inhomogeneous systems whereas the variational methods are unimpaired. In order to keep closer to the physical variables we shall not follow Antonov's trail but shall make our own. We thereby discover a simplified form of Antonov's condition. Our method gains a close analogy with gaseous stability problems and a stimulating similarity to Schrödinger's equation.

The systems to which this work applies are

(i) Those whose distribution functions depend on energy[3] (in rotating axes) only.

(ii) More general systems whose other integrals (other than the energy) are not disturbed by the class of oscillations contemplated. An example will elucidate what we mean. An axially symmetrical system may have a distribution function of the form $F = F(E, \varpi_z)$ where

$$E = c^2/2 - \Psi, \qquad \varpi_z = (\mathbf{r} \times \mathbf{c})_z,$$

$$\mathbf{r} = (x, y, z), \qquad \mathbf{c} = (u, v, w),$$

and Ψ is the gravitational potential.

If we only contemplate axially symmetrical disturbances of such a system then the angular momentum of each star will be undisturbed in the sense used above. A sufficient condition for the stability of such a system against all axially symmetrical perturbations is derived below.

Our approach to the problem is through the linearized Boltzmann-Liouville equation for the total mass distribution function f_T. $f_T(\mathbf{r}, \mathbf{c}, t) \, d^3r \, d^3c$ is then the total mass in a small box of volume d^3r about $\mathbf{r}$ moving with velocity in a range d^3c about $\mathbf{c}$, at time t. The equation is:

---

[3] Here and hereafter energy will normally mean energy per unit mass.

(26) $\quad \dfrac{\partial f_T}{\partial t} + \mathbf{c} \cdot \dfrac{\partial f_T}{\partial \mathbf{r}} + \left(\dfrac{\partial \Psi_T}{\partial \mathbf{r}} - 2\Omega \times \mathbf{c} + \Omega^2 \mathbf{R} \right) \cdot \dfrac{\partial f_T}{\partial \mathbf{c}} = 0,$

where $\mathbf{R} = (x, y, 0)$; $\Omega = (0, 0, \Omega)$ is the angular velocity of our axes, and Ψ_T is the total gravitational potential.

We shall use F and Ψ for the unperturbed mass distribution function and potential of the equilibrium. We also define an operator D by:

(27) $\qquad D = \mathbf{c} \cdot \partial/\partial \mathbf{r} + (\partial \Psi/\partial \mathbf{r} - 2\Omega \times \mathbf{c} + \Omega^2 \mathbf{R}) \cdot \partial/\partial \mathbf{c}.$

F satisfies the steady-state Boltzmann-Liouville equation which may be written

(28) $\qquad\qquad\qquad DF = 0.$

If we call the perturbation in mass distribution function and potential f and ψ respectively, then $f_T - F = f$, $\Psi_T - \Psi = \psi$.

Subtracting Equation (28) from Equation (26) and linearizing in the perturbed quantities we obtain

(29) $\qquad\qquad \partial f/\partial t + Df + (\partial \psi/\partial \mathbf{r}) \cdot (\partial F/\partial \mathbf{c}) = 0.$

If the perturbation in the gravity field arises from the perturbation in the stellar density then Poisson's equation reads

(30) $\qquad\qquad \nabla^2 \psi = -4\pi G \displaystyle\int f d^3 c,$

where G is the constant of gravitation and it is understood that (30) should be solved subject to the boundary condition that $\psi = O(1/R)$ at ∞.[4]

In performing stability analyses it has been customary to use f as the dependent variable and to work from Equations (29) and (30). However, such an approach has disadvantages. f gives the excess mass of stars at a given point and given velocity, in the perturbed system, over the mass in the unperturbed system. The stars that are being compared have energies $(c^2/2) - \psi - \Psi$ and $(c^2/2) - \Psi$, respectively. The great importance of energy in determining stability might therefore lead us to compare systems by comparing the total mass of stars with the same *energy* at the same point rather than with the same velocity at the same point.

—————————

[4] Actually $O(1/R^2)$ since the total mass does not change in perturbation.

To find the quantity q that expresses perturbations in this way we must first decide how the unperturbed distribution function F depends on energy. We write

$$(31) \qquad E = c^2/2 - \Psi - \Omega^2 R^2/2.$$

$D(E) = 0$, so $F = F(E)$ is a possible solution of Equation (28). However, in general there are other isolating integrals independent of the energy. The general solution of (28) is thus $F = F(E, I_2, I_3, \cdots)$, where $I_1 = E$ and the I_i are independent isolating first integrals of the equations of motion of a star moving in the unperturbed system (which do not involve t explicitly). We shall assume that F depends on known integrals I_i which are chosen for their simplicity as functions of $\mathbf{r}$ and $\mathbf{c}$. With the I_i known we can consider the distribution function of any steady-state system to be a definite, known function of E, I_2, I_3, etc.; $\partial F/\partial E$ is therefore well defined. Whereas

$$(32) \qquad \begin{aligned} f(\mathbf{r}, \mathbf{c}, t) &= f_T(\mathbf{r}, \mathbf{c}, t) - F(E, I_2, \cdots) \\ &= f_T(\mathbf{r}, \mathbf{c}, t) - F(c^2/2 - \Psi - \Omega^2 R^2/2, I_2, \cdots), \end{aligned}$$

we define q by

$$\begin{aligned} q(\mathbf{r}, \mathbf{c}, t) &= f_T(\mathbf{r}, \mathbf{c}, t) - F(c^2/2 - \Psi_T - \Omega^2 R^2/2, I_2, \cdots) \\ &= f_T(\mathbf{r}, \mathbf{c}, t) - F(E - \psi, I_2, \cdots) \approx f + \psi \, \partial F/\partial E, \end{aligned}$$

where we have linearised in the last equation.

Henceforth we shall write F' for $\partial F/\partial E$. Since F' is a function of integrals of the motion which are time independent,

$$(33) \qquad D(F') = 0.$$

Substituting for f in terms of q in Equation (29) we obtain

$$(34) \qquad \frac{\partial q}{\partial t} + Dq = F' \frac{\partial \psi}{\partial t} + F' D\psi - \frac{\partial \psi}{\partial \mathbf{r}} \cdot \frac{\partial F}{\partial \mathbf{c}}.$$

But

$$\frac{\partial F}{\partial \mathbf{c}} = \sum_{i=1}^{\cdots} \frac{\partial F}{\partial I_i} \frac{\partial I_i}{\partial \mathbf{c}} = F' \mathbf{c} + \sum_{i=2}^{\cdots} \frac{\partial F}{\partial I_i} \frac{\partial I_i}{\partial \mathbf{c}}$$

and $D\psi = \mathbf{c} \cdot \partial \psi/\partial \mathbf{r}$, hence the Boltzmann-Liouville Equation (34) for q can be written

$$(35) \qquad \frac{\partial q}{\partial t} + Dq = F' \frac{\partial \psi}{\partial t} + \left(\sum_{i=2}^{\cdots} \frac{\partial F}{\partial I_i} \frac{\partial I_i}{\partial \mathbf{c}} \right) \cdot \frac{\partial \psi}{\partial \mathbf{r}}.$$

We shall need to supplement Equation (35) with Poisson's equation, written in terms of q,

$$\nabla^2\psi = -4\pi G \int (q - F'\psi)\,d^3c,$$

hence defining the operator S by

(36) $$S = -\frac{1}{4\pi G}\nabla^2 + \int F'd^3c$$

we have

(37) $$S\psi = \int q\,d^3c.$$

There is an analogy between S and the Hamiltonian operator for a single particle moving in a potential well in quantum mechanics. There one has

(38) $$H = -\hbar^2\nabla^2/2m + V(\mathbf{r}).$$

We shall confine ourselves to systems for which $F' \le 0$—fewer stars at higher speeds when the other integrals are kept fixed. In these circumstances the analogous $V(r)$ is negative, showing that we have the Hamiltonian of an attractive potential.

So far we have been perfectly general. In what follows we restrict ourselves to situations in which

(39) $$\sum_{i=2}^{\cdots} \frac{\partial F}{\partial I_i}\frac{\partial I_i}{\partial \mathbf{c}}\cdot\frac{\partial \psi}{\partial \mathbf{r}} = 0.$$

The most obvious class of these is the class of systems whose distribution functions are functions of E alone (for some value of Ω, possibly zero). In such a case all the $\partial F/\partial I_i$, $i \ge 2$, are zero. A second, important class is that of the axially symmetrical disturbances of axially symmetrical stellar systems whose distribution functions depend on energy and angular momentum only. Then

$$F = F(E, \varpi_z) = F(E, Rc_\phi),$$

so

$$\frac{\partial F}{\partial \varpi_z}\frac{\partial \varpi_z}{\partial \mathbf{c}}\cdot\frac{\partial \psi}{\partial \mathbf{r}} = \frac{\partial F}{\partial \varpi_z}R\hat{\phi}\cdot\frac{\partial \psi}{\partial \mathbf{r}} = 0$$

by reason of the axial symmetry of ψ. Here $\hat{\phi}$ is the unit toroidal vector right handed about O_3 and $c\phi = \mathbf{c}\cdot\hat{\phi}$.

A third class consists of the spherically symmetrical oscillations of a nonrotating stellar system whose distribution function depends on E and $\varpi^2 = (\mathbf{r} \times \mathbf{c})^2$ only, so $F = F(E, \varpi^2)$. There are numerous other systems which can have oscillations preserving the symmetry associated with an integral. All such oscillations can be discussed by this method. ·

A.2. *A Sufficient Condition for Stability.* Multiply Equation (35) by $q/-F'$ and integrate over all position and velocity space:

$$\int \frac{q}{-F'} \frac{\partial q}{\partial t} d^6\tau + \int \frac{q\, Dq}{-F'} d^6\tau = -\int q \frac{\partial \psi}{\partial t} d^6\tau$$

where we have used Equation (39).

Hence using Equation (37)

$$\frac{\partial}{\partial t} \left(\frac{1}{2} \int \frac{q^2}{-F'} d^6\tau \right) + \int \left\{ \int qd^3c \frac{\partial}{\partial t} \left(S^{-1} \int qd^3c \right) \right\} d^3r$$

$$= \int \frac{q\, Dq}{F'} d^6\tau$$

or, since S is Hermitian,

$$(40) \qquad \frac{\partial}{\partial t} \left[\frac{1}{2} \int \frac{q^2}{-F'} d^6\tau + \frac{1}{2} \int \left\{ \int qd^3c\, S^{-1} \int qd^3c \right\} d^3r \right]$$

$$= \int \frac{q\, Dq}{F'} d^6\tau.$$

We now show that D is antihermitian, so that the right-hand side is zero. We write

$$(41) \qquad\qquad \mathbf{K} = (\partial \Psi / \partial \mathbf{r}) - 2\Omega \times \mathbf{c} + \Omega^2 \mathbf{R}$$

and note that $(\partial/\partial \mathbf{c}) \cdot (\mathbf{K}f) = \mathbf{K} \cdot (\partial f/\partial \mathbf{c})$ for any f; consequently

$$Df = \left(\mathbf{c} \cdot \frac{\partial}{\partial \mathbf{r}} + \mathbf{K} \cdot \frac{\partial}{\partial \mathbf{c}} \right) f = \frac{\partial}{\partial \mathbf{r}} \cdot (f\mathbf{c}) + \frac{\partial}{\partial \mathbf{c}} \cdot (f\mathbf{K}).$$

Now for functions f and g that vanish sufficiently far from the cluster and which are zero wherever in phase space $E > 0$,

$$\int D(fg) d^6\tau = \int fg\mathbf{c}d^3c \cdot dS + \int fg\mathbf{K}d^3r \cdot dS_c = 0.$$

So

$$0 = \int fDg d^6\tau + \int gDf d^6\tau,$$

which shows that D is antihermitian.

On the right of Equation (40) we have

$$\int \frac{q}{F'} Dq d^6\tau = -\int qD\left(\frac{q}{F'}\right) d^6\tau = -\int \frac{q}{F'} Dq d^6\tau,$$

therefore the left-hand side of Equation (40) vanishes. Thus

(42) $$\frac{1}{2}\int \frac{q^2}{-F'} d^6\tau + \frac{1}{2}\int \left\{ \int qd^3cS^{-1}\int qd^3c \right\} d^3r - \text{const.}$$

If S is positive definite then each term is positive (provided $F' < 0$). Hence if the whole disturbance starts small each term will remain less than the small constant on the right. Hence a disturbance that starts small remains small, showing stability. Thus if S is positive definite there are no instabilities.

A.3. *Connection with the Energy.* The energy of the perturbation is

$$\delta W = \frac{1}{2}\int (f_T - F)\left[(c_\phi + \Omega R)^2 + c_R^2 + w^2\right] d^6\tau$$

$$- \frac{G}{2}\int \frac{\int f_T d^3c \int f_T^* d^3c^*}{|\mathbf{r} - \mathbf{r}^*|} d^3r d^3r^*$$

$$+ \frac{G}{2}\int \frac{\int F d^3c \int F^* d^3c^*}{|\mathbf{r} - \mathbf{r}^*|} d^3r d^3r^*$$

$$= \frac{1}{2}\int fc^2 d^6\tau + \Omega \int f\left(Rc_\phi + \frac{\Omega}{2}R^2\right) d^6\tau$$

$$- 2\int \frac{G}{2} \frac{\int f d^3c \int F^* d^3c^*}{|\mathbf{r} - \mathbf{r}^*|} d^3r d^3r^*$$

$$- \frac{G}{2}\int \frac{\int f d^3c \int f^* d^3c^*}{|\mathbf{r} - \mathbf{r}^*|} d^3r d^3r^*$$

$$= \int f\left(\frac{c^2}{2} - \Psi\right) d^6\tau + \Omega \int \left(Rc_\phi + \frac{\Omega}{2}R^2\right) d^6\tau$$

$$- \frac{G}{2}\int \frac{\int f d^3c \int f^* d^3c^*}{|\mathbf{r} - \mathbf{r}^*|} d^3r d^3r^*.$$

Also, the angular momentum of the perturbation is

$$\delta H = \int f(c_\phi + \Omega R)\, R d^6\tau,$$

so

(43)
$$\begin{aligned}
\delta W &= \int f\left(\frac{c^2}{2} - \Psi - \frac{1}{2}\Omega^2 R^2\right) d^6\tau + \Omega\delta H \\
&\quad - \frac{G}{2}\int \frac{\int f d^3c \int f^* d^3c^*}{|\mathbf{r} - \mathbf{r}^*|}\, d^3r d^3r^* \\
&= \int fE d^6\tau + \Omega\delta H - \frac{G}{2}\int \frac{\int f d^3c \int f^* d^3c^*}{|\mathbf{r} - \mathbf{r}^*|}\, d^3r d^3r^*.
\end{aligned}$$

To transform the first term into a quadratic form in f we use a trick due to Newcomb. Since f_T remains constant as we follow the flow, $\int J(f_T)\, d^6\tau$ is a constant of the motion for an arbitrary function J. Further, when we are dealing with systems in which integrals other than the energy appear, but are conserved for the motions contemplated, we can generalise this to

$$\int J(f_T, I_i \cdots)\, d^6\tau = \text{const}$$

for arbitrary J. Subtracting the unperturbed value we obtain

$$\int \{ J(f_T, I_i \cdots) - J(F, I_i \cdots) \}\, d^6\tau = \text{const} \qquad \text{(small)};$$

so correct to second order in f

(44)
$$\int \left[f\frac{\partial J}{\partial F} + \frac{1}{2}f^2\frac{\partial^2 J}{\partial F^2} \right] d^6\tau = \text{const},$$

where $\partial J/\partial F$, $\partial^2 J/\partial F^2$ are to be evaluated at $(E, I_i \cdots)$. Now J is a function $F(E, I_i \cdots)$, $I_i \cdots$, so it may be regarded as a function of $I, I_i \cdots$ which is still arbitrary; we choose

$$J = \int^E \left(\frac{\partial F}{\partial E'}\right)_{I_i} E' dE'.$$

Then

$$\frac{\partial J}{\partial F} = \frac{\partial J}{\partial E}\left(\frac{\partial E}{\partial F}\right)_{I_i} = \left(\frac{\partial F}{\partial E}\right)_{I_i}\left(\frac{\partial E}{\partial F}\right)_{I_i} E = E$$

and

$$\frac{\partial^2 J}{\partial F^2} = \left(\frac{\partial E}{\partial F}\right)_{I_i} = \left\{\frac{\partial F}{\partial E}\right\}_{I_i}^{-1} = \frac{1}{F'}.$$

Hence from (44)

$$\int fE d^6\tau = +\frac{1}{2}\int \frac{f^2}{-F'} d^6\tau + \text{const.}$$

Returning to our expression for the energy and remembering that $q = f + F'\psi$,

$$\delta W - \Omega\delta H + \text{const} = \frac{1}{2}\int \frac{f^2}{-F'} d^6\tau - \frac{1}{2}\int f\psi d^6\tau$$

$$= \frac{1}{2}\int \frac{q^2}{-F'} d^3c + \frac{1}{2}\int\left[\int qd^3c S^{-1}\int qd^3c\right] d^3r,$$

which is the expression (42) derived earlier.

The potential of the Schrödinger operator is $\int (\partial F/\partial E) d^3c$. But $\rho_0 = \int F(E, \varpi_z) d^3c$; so for axially symmetrical systems and those for which $F = F(E)$

$$(45) \qquad\qquad \left(\frac{\partial \rho_0}{\partial \Psi}\right)_R = -\int \frac{\partial F}{\partial E} d^3c,$$

which demonstrates that only the density distribution enters the Schrödinger operator.

B. Stability of gaseous systems.[5]

B.1. *A Sufficient Condition for Stability.* The equations of the perturbed motion of a barytropic gas rotating uniformly in its unperturbed state are

$$(46) \qquad \frac{\partial \mathbf{u}}{\partial t} + 2\Omega \times \mathbf{u} = \delta\left(-\frac{1}{\rho}\nabla p\right) + \nabla\psi_1 = \nabla\left(-\frac{p_0'}{\rho_0}\rho_1 + \psi_1\right),$$

where $\mathbf{u}$ is the perturbed velocity in the rotating axes, $p_0' = dp_0/d\rho_0$, and noughts denote equilibrium values and ones perturbed values. The perturbed continuity equation reads

$$(47) \qquad\qquad \partial\rho_1/\partial t + \text{div}(\rho_0\mathbf{u}) = 0.$$

We write

$$(48) \qquad\qquad Q = \rho_1 - (\rho_0/p_0')\,\psi_1.$$

---

[5] See [16] and [17].

Then

(49) $$-4\pi G\rho_1 = -4\pi GQ - 4\pi G(\rho_0/p_0')\psi_1,$$

so Poisson's equation reads

(50) $$((-4\pi G)^{-1}\nabla^2 - \rho_0/p_0')\psi_1 = Q.$$

From the unperturbed equation of equilibrium along O_z,

(51) $$\partial\Psi/\partial z = (1/\rho_0)(\partial p_0/\partial z) = (p_0'/\rho_0)(\partial\rho_0/\partial z),$$

we get

$$\rho_0/p_0' = (\partial\rho_0/\partial\Psi)_R.$$

Thus the potential in the Schrödinger operator of Equation (50) is the same as that found in Equation (45) for a stellar system. We are therefore justified in writing for Equation (50)

$$S\psi = Q.$$

We now write the continuity equation in terms of Q to obtain

(52) $$\frac{\partial Q}{\partial t} + \frac{\rho_0}{p_0'}\frac{\partial\psi}{\partial t} + \text{div}(\rho_0\mathbf{u}) = 0.$$

Still following the stellar dynamical case we multiply this continuity equation by $(p_0'/\rho_0)Q$ and integrate over all space:

(53) $$\int\frac{p_0'}{\rho_0}Q\frac{\partial Q}{\partial t}d^3r + \int Q\frac{\partial\psi}{\partial t}d^3r + \int\frac{p_0'}{\rho_0}Q\,\text{div}(\rho_0\mathbf{u})d^3r = 0.$$

Noting that the surface integral $\int p_0'Q\mathbf{u}\cdot d\mathbf{S}$ over a large sphere will vanish we write the last term as

$$-\int\nabla\left(\frac{p_0'}{\rho_0}Q\right)\cdot\rho_0\mathbf{u}d^3r.$$

However the equation of motion when written in terms of Q reads

(54) $$\frac{\partial u}{\partial t} + 2\Omega\times\mathbf{u} = -\nabla\left(\frac{p_0'}{\rho_0}Q\right),$$

hence instead of Equation (28) we may write

(55) $$\frac{\partial}{\partial t}\left(\frac{1}{2}\int\frac{p_0'}{\rho_0}Q^2d^3r\right. + \int Q\frac{\partial\psi}{\partial t}d^3r$$
$$+ \int\left(\frac{\partial\mathbf{u}}{\partial t} + 2\Omega\times\mathbf{u}\right)\cdot\rho_0\mathbf{u}d^3r = 0,$$

which by reason of Equation (51) we may write as

$$(56) \quad \frac{\partial}{\partial t}\left[\frac{1}{2}\int\frac{Q^2}{\rho_0/p_0'}d^3r + \frac{1}{2}\int\rho_0 u^2 d^3r + \frac{1}{2}\int QS^{-1}Qd^3r\right] = 0.$$

The "extra" term $(1/2)\int\rho_0 u^2 d^3r$ is not really extra because when one compares Equations (42) and (56) one sees that q contains velocity perturbations whereas Q does not.

If one writes

$$q_1 = \tfrac{1}{2}\{q(\mathbf{r},\mathbf{c},t) + q(\mathbf{r}, -\mathbf{c},t)\},$$

$$q_2 = \tfrac{1}{2}\{q(\mathbf{r},\mathbf{c},t) - q(\mathbf{r}, -\mathbf{c},t)\},$$

then q_2 does not contribute to $\int q d^3 c$.

Equation (42) may now be written

$$\frac{1}{2}\int\frac{q_1^2}{-F'}d^6\tau + \frac{1}{2}\int\frac{q_2^2}{-F'}d^6\tau$$

$$+ \frac{1}{2}\int\left\{\int q d^3 c S^{-1}\int q d^3 c\right\}d^3 r = \text{const},$$

in which a velocity term (q_2 is antisymmetric in velocity) has appeared. Each term being positive if S is positive definite, we have a sufficient condition for stability.

B.2. *Necessity of the Condition for Nonrotating Systems* From Equation (46) with $\Omega = 0$

$$(57) \quad\quad\quad\quad i\omega\mathbf{u} = \nabla\left((p_0'/\rho_0)\,Q\right).$$

Using Equation (52) we obtain

$$-\omega^2(Q + (\rho_0/p_0')\,\psi) = -\operatorname{div}\left[\rho_0\nabla\,(p_0'Q/\rho_0)\right]$$

or, multiplying by p_0'/ρ_0 and using (51),

$$(58) \quad\quad\quad\quad + \omega^2 BQ = CQ.$$

Define operators B and C by

$$BQ = p_0'Q/\rho_0 + S^{-1}Q,$$

$$CQ = (p_0'/\rho_0)\operatorname{div}\left[\rho_0\nabla\,(p_0'Q/\rho_0)\right].$$

Then B and C are Hermitian since

$$\int P^*BQd^3r = \int P^*Q\frac{p_0'}{\rho_0}d^3r + \int P^*S^{-1}Qd^3r$$

$$= \left[\int Q^*P\frac{p_0'}{\rho_0}d^3r + \int Q^*S^{-1}Pd^3r\right]^*$$

$$= \left[\int Q^*BPd^3r\right]^*$$

and also

$$-\int P^*\frac{p_0'}{\rho_0}\operatorname{div}\left[\rho_0\nabla\left(\frac{p_0'Q}{\rho_0}\right)\right]d^3r$$

$$= +\int \rho_0\nabla\left(P^*\frac{p_0'}{\rho_0}\right)\cdot\nabla\left(Q\frac{p_0'}{\rho_0}\right)d^3r$$

$$= -\left\{\int Q^*\frac{p_0'}{\rho_0}\operatorname{div}\left[\rho_0\nabla\left(\frac{p_0'Q}{\rho_0}\right)\right]d^3r\right\}^*.$$

Note also that C is positive definite unless there is a stationary mode, because by Equation (57)

$$\int Q^*CQd^3r = +\int \rho_0\nabla\left(Q^*\frac{p_0'}{\rho_0}\right)\cdot\nabla\left(Q\frac{p_0'}{\rho_0}\right)d^3r > 0$$

it can only equal zero when Qp_0'/ρ_0 is constant, which from Equation (54) with $\Omega = 0$ means a stationary velocity field. We can without great loss of generality consider systems that are not marginally stable so that C may be taken to be positive definite.

Let us return to Equation (58). Since C is positive definite the set of Q's that simultaneously diagonalise B and C will be a complete set, and the eigen-Q's corresponding to different eigenvalues will be B-orthogonal; those with the same ω can be orthogonalised.

Suppose there exists a Q (not necessarily an eigen-Q) such that $\int Q^*BQd^3r < 0$, then expanding Q as

$$Q = \sum_\omega q_\omega Q_\omega$$

we have by the B-orthogonality of the Q

$$\int Q^*BQd^3r = \sum_\omega (q_\omega)^2\int Q_\omega^*BQ_\omega d^3r < 0.$$

Hence for at least one ω, $\int Q_\omega^*BQ_\omega d^3r < 0$. But from Equation (58) $\omega^2\int Q_\omega^*BQ_\omega d^3c = \int Q_\omega^*CQ_\omega d^3c > 0$, so $\omega^2 < 0$ for at least one ω.

Thus to prove instability we need only show that there is a Q such that $\int Q^*BQd^3r < 0$.

We now show that if there exists a ψ such that $\int \psi^*S\psi d^3r < 0$, then there is a Q such that $\int Q^*BQd^3r < 0$.

Consider the eigenvalue equation

(59) $S\psi = \lambda \left(\rho_0/p_0'\right) \psi.$

Since ρ_0/p_0' is positive the eigenfunctions will form a complete set, and by a similar argument to the above there will therefore be a negative eigenvalue λ and a corresponding eigenfunction ψ; hence

$$\left(-\frac{\nabla^2}{4\pi G} - \frac{\rho_0}{p_0'}\right)\psi = \lambda \frac{\rho_0}{p_0'}\psi,$$

$$\int \frac{\nabla \psi^* \cdot \nabla \psi}{4\pi G}d^3r = (\lambda + 1)\int \psi^* \frac{\rho_0}{p_0'}\psi d^3r;$$

so although λ is negative, $\lambda + 1$ is positive. Now consider the Q's given by $Q = \lambda \left(\rho_0/p_0'\right)\psi$.

(60) $\int Q^*BQd^3r = \lambda^2 \int \psi^* \frac{\rho_0}{p_0'}\psi d^3r + \lambda \int \psi^* \frac{\rho_0}{p_0'}\psi d^3r,$

since by construction $\psi = S^{-1}Q$ from Equations (59) and (60). Hence

$$\int Q^*BQd^3r = \lambda(\lambda + 1)\int \psi^* \frac{\rho_0}{p_0'}\psi d^3r < 0 \text{ since } \lambda(\lambda + 1) < 0.$$

Summing up we have proved that if there is a ψ such that $\int \psi^*S\psi d^3r < 0$, then there is a Q such that $\int Q^*BQd^3r < 0$ and an unstable normal mode with $\omega^2 < 0$. Thus if there is a bound state for the associated Schrödinger equation, then the system is unstable. The nature of the instability is easily guessed at from the associated bound state of the Schrödinger equation; it is in some sense the formation of the bound subsystem.

C. Antonov's necessary and sufficient variational principle.

C.1. *A Sufficient Condition for Stability when the Unperturbed Distribution Function is Dependent on Energy only.* The perturbed Boltzmann-Liouville equation reads

(61) $\frac{\partial f}{\partial t} + Df + \frac{\partial \psi}{\partial \mathbf{r}} \cdot \frac{\partial F}{\partial \mathbf{c}} = 0,$

where

$$D = \mathbf{c} \cdot \frac{\partial}{\partial \mathbf{r}} + \frac{\partial \Psi}{\partial \mathbf{r}} \cdot \frac{\partial}{\partial \mathbf{c}},$$

$$\frac{\partial F}{\partial \mathbf{c}} = \frac{\partial F}{\partial E} \mathbf{c} = \mathbf{c} F'.$$

Note

$$D\psi = \mathbf{c} \cdot \partial\psi/\partial\mathbf{r},$$

and

$$D(F') = F'' D(E) = 0;$$

as before, D is anti-self-adjoint.

From the self-gravitation equation,

$$(62) \qquad \psi(\mathbf{r}) = G \int \frac{f(r', c')}{|\mathbf{r} - \mathbf{r}'|} \, d^6\tau'.$$

We now divide f into symmetric and antisymmetric parts:

$$(63) \qquad \begin{aligned} f_1 &= \tfrac{1}{2} \{ f(\mathbf{r}, \mathbf{c}, t) + f(\mathbf{r}, -\mathbf{c}, t) \}, \\ f_2 &= \tfrac{1}{2} \{ f(\mathbf{r}, \mathbf{c}, t) - f(\mathbf{r}, -\mathbf{c}, t) \}. \end{aligned}$$

Then we have

$$(64) \qquad \dot{f}_1 + Df_2 = 0,$$

$$(65) \qquad \dot{f}_2 + Df_1 + F'D\psi = 0.$$

Hence

$$(66) \qquad \dot{f}_2 = - Df_1 - F'D \left(G \int \frac{f_1' d^6\tau'}{|\mathbf{r} - \mathbf{r}'|} \right).$$

If we define

$$Af = \frac{D^2}{+F'} f + GD \int \frac{D'f'}{|r - r'|} \, d^6\tau'$$

and then replace $\dot{f}_1$ by $-Df_2$, we have

$$(67) \qquad \frac{1}{F'} \cdot -\frac{\partial^2 f_2}{\partial t^2} = Af_2.$$

A is self-adjoint because (since D is anti-self-adjoint)

$$\int g A f d^6\tau = - \int \frac{Dg\,Df}{-F'} \, d^6\tau - G \iint \frac{Dg\,D'f'}{|\mathbf{r} - \mathbf{r}'|} \, d^6\tau \, d^6\tau'$$

$$= \int f A g d^6\tau.$$

Multiply Equation (67) by $\dot{f}_2$ and integrate over all space to obtain

$$\frac{\partial}{\partial t}\left[\frac{1}{2}\int\left(f_2 A f_2 + \frac{\dot{f}_2^2}{-F'}\right)d^6\tau\right] = 0.$$

So

$$\frac{1}{2}\int\left(f_2 A f_2 + \frac{\dot{f}_2^2}{-F'}\right)d^6\tau = \text{const.}$$

If f_2 is small at time $t = 0$ the constant is small. If A is positive definite and $F' < 0$, then the left-hand side is the sum of positive terms, and each must therefore remain small. Hence f_2, $\dot{f}_2$ must remain small, so by (64) and (65) f_1 and therefore f must remain small. Hence we have stability. Thus for clusters with $F' < 0$ a sufficient condition for stability is that A be positive definite.

C.2. *Necessity of this Condition for Clusters with* $F' < 0$. We look for normal modes, so Equation (67) reduces to

(68) $(\omega^2/-F')f_2^{(\omega)} = +Af_2^{(\omega)},$

where the normal mode has an $e^{i\omega t}$-dependence. From (68)

$$\omega^2 = \frac{\int f_2^{(\omega)*}Af_2^{(\omega)}d^6\tau}{\int (f_2^{(\omega)*}f_2^{(\omega)}/-F')d^6\tau},$$

which is real. Also, the eigenfunctions are orthogonal; for

(69) $\omega_1^2\int\frac{f_2^{\omega 1*}f_2^{\omega 2}}{-F'}d^6\tau = \int f_2^{\omega 1*}Af_2^{\omega 2}d^6\tau.$

Taking complex conjugates

(70) $\omega_1^2\int\frac{f_2^{\omega 2*}f_2^{\omega 1}}{-F'}d^6\tau = \int f_2^{\omega 2*}Af_2^{\omega 1}d^6\tau.$

We now write ω_2 for ω_1 and ω_1 for ω_2 in (69) and subtract (70) to obtain

$$(\omega_2^2 - \omega_1^2)\int\frac{f_2^{\omega 2*}f_2^{\omega 1}}{-F'}d^6\tau = 0,$$

which proves orthogonality of eigenfunctions with different eigenvalues.

To prove the necessity of the condition that A be positive, we assume that there exists a function f_2 such that

(71)
$$\int f_2^* A f_2 d^6\tau < 0,$$

and attempt to show that there must be an instability, that is, a normal mode with negative ω^2. By the expansion theorem we may write

$$f_2 = \sum_\omega c_\omega f_2^{(\omega)}.$$

so

$$\int f_2 A f_2 d^6\tau = \sum_\omega \sum_{\omega'} c_\omega^* c_{\omega'} \int f_2^{\omega*} A f_2^{\omega'} d^6\tau$$

$$= \sum_\omega \omega^2 c_\omega^* c_\omega \int \frac{f_2^{\omega*} f_2^\omega}{-F'} d^6\tau.$$

By (71) this sum is less than zero; but if all ω^2 were $\geqq 0 \cdot$ all terms would be nonnegative. Hence at least one ω^2 must be negative. Hence we have instability. Thus if there exists a function f_2 such that $\int f_2 A f_2 d^6\tau < 0$, the system is unstable. This completes the proof of Antonov's energy principle.

C.3. *Reduction of Antonov's Principle.*

$$\int f_2^* A f_2 d^6\tau = \int \frac{D f_2^* D f_2}{-F'} d^6\tau - G \int \int \frac{D f_2^* D' f_2^*}{|\mathbf{r} - \mathbf{r}'|} d^6\tau \, d^6\tau.$$

We write $D f_2 = \eta$, a symmetrical function of $\mathbf{c}$, and define the operator B by

$$B\eta = \frac{\eta}{-F'} - G \int \frac{\eta'}{|\mathbf{r} - \mathbf{r}'|} d^6\tau'.$$

Then

(72)
$$\int f_2^* A f_2 d^6\tau = \int \eta B\eta d^6\tau.$$

Obviously a sufficient condition for stability is $\int \eta^* B\eta d^6\tau > 0$ for all symmetrical functions η obeying the boundary conditions; but this is still not a necessary condition since not all functions η can be expressed as the D-maps of acceptable functions f_2.

We now make a connection with our energy principle by proving that the necessary and sufficient condition that B should be positive definite is that our Schrödinger operator S should be.

Put

(73) $$p = \eta + F'\psi,$$

where

$$\psi = G \int \frac{\eta'}{|\mathbf{r} - \mathbf{r}'|} \, d^6\tau'.$$

Then

$$\nabla^2\psi = -4\pi G \int \eta \, d^3c$$

and

(74) $$\int p \, d^3c = \left(\frac{\nabla^2}{-4\pi G} + \int F' d^3c \right) \psi = S\psi,$$

where S is our Schrödinger operator. It is simple to show that

(75) $$\int \eta^* B\eta \, d^6\tau = \int \frac{p^*p}{-F'} d^6\tau + \int \left\{ \int p^* d^3c \, S^{-1} \int p \, d^3c \right\} d^3r,$$

where $S^{-1} \int p \, d^3c = \psi$ from (74). It is clear from (75) that, since $F' < 0$, B is positive definite if S is. Hence a sufficient condition for stability is that S is positive definite. (Then B is and so A is.)

Consider now the case when we know some function obeying the boundary conditions such that $\int \psi^* S\psi \, d^3r < 0$. We shall produce a function p such that the expression (75) is negative. The η may be reconstructed from the p by the formula

(76) $$\eta = p - F'\psi = p - F' \left(S^{-1} \int p \, d^3c \right).$$

Consider the eigenvalue equation

(77) $$S\psi_\lambda = \lambda \int - F' d^3c \psi_\lambda.$$

Since $F' < 0$ the right-hand side is a positive definite operator acting on ψ_λ; it may therefore be simultaneously diagonalised with S, and the eigenfunctions will form a complete set. Expand the given ψ that makes $\int \psi^* S\psi \, d^3r < 0$ in terms of these ψ_λ. By a similar argument to that given under Equation (71) we may prove one eigenvalue λ to be negative. Consider the corresponding ψ_λ. Then

$$\lambda \int \psi_\lambda^* \int F' d^3c \psi_\lambda \, d^3r = \int \psi_\lambda^* S\psi_\lambda \, d^3r = \int \frac{\nabla \psi_\lambda^* \cdot \nabla \psi_\lambda}{4\pi G} \, d^3r$$

$$- \int \psi_\lambda^* \int - F' d^3c \psi_\lambda \, d^3r,$$

where the definition of S has been used.

Hence

$$(\lambda + 1) \int \psi_\lambda \int - F' d^3 c \psi_\lambda d^3 r = \int \frac{\nabla \psi_\lambda^* \cdot \nabla \psi_\lambda}{4\pi G} d^3 r > 0.$$

So $(\lambda + 1) > 0$ although we chose λ to be negative, hence

(78) $$\lambda(\lambda + 1) < 0.$$

Now choose

(79) $$p = \lambda F' \psi_\lambda,$$

then the ψ corresponding to that p is the solution of $(\int p d^3 c) = S\psi$, which is just ψ_λ by Equation (77). Substituting the expression (79) for p into the right-hand side of Equation (75) we have

$$\int \eta^* B \eta d^6 \tau = (\lambda^2 + \lambda) \int \psi_\lambda^* \int - F' d^3 c \psi d^3 r < 0$$

from Equation (78). Also, from Equation (76), the corresponding η is

$$\eta = \lambda F' \psi_\lambda - F' \psi_\lambda = (\lambda - 1) F' \psi_\lambda,$$

which obeys the boundary conditions.

The only outstanding question is whether an η of this form is the D-map of an acceptable function f_2. If it is, then the Schrödinger operator condition is necessary and sufficient, and is equivalent to Antonov's. If it is not, then the Schrödinger operator condition is a sufficient condition which is not necessary for stability.

Fundamental References

1. W. B. Thompson, *An introduction to plasma physics*, Pergamon Press, New York, 1962.

2. T. H. Stix, *The theory of plasma waves*.

3. L. D. Landau, Acad. Sci. USSR J. Phys. **10** (1946), 25-34.

4. O. Penrose, Phys. Fluids 3 (1960), 258.

5. I. B. Bernstein, Phys. Rev. **109** (1958), 10.

References Concerned with the Stellar Dynamical Case

6. V. A. Antonov, Astronom. Ž. 37 (1960), 918-926 = Soviet Astronom. AJ 4 (1961), 859-867.

7. ———, Vestnik Leningrad. Univ. Ser. Mat. Meh. Astronom. 19 (1962), 96.

8. D. Lynden-Bell, *Theory of orbits in the solar system and in stellar systems*, I.A.U. Sympos. no. 25, Thessaloniki, 1964, Academic Press, New York, 1966, p. 78.

9. ———, Monthly Notices Roy. Astronom. Soc. **124** (1962), 95-123.

10. R. Simon, Bull. Acad. Roy. Belg. Ser. **47** (1962), 7.
11. P. A. Sweet, Monthly Notices Roy. Astronom. Soc. **125** (1963), 285.
12. A. Toomre, Astrophys. J. **139** (1964), 1217.
13. W. B. Wilson and D. Lynden-Bell, Unpublished.

General References

14. R. L. Liboff, Phys. Lett. **3** (1963), 322.
15. W. B. Thompson, Private communication.
16. F. J. Dyson, preprint 1964.
17. R. C. Tolman, Astrophys. J. **90** (1939), 541.

References added 1974

P. Bartholomew, Monthly Notices Roy. Astronom. Soc. **151** (1971), 333.

A. J. Kalnajs, Astrophys. J. **166** (1971), 275.

D. Lynden-Bell, Monthly Notices Roy. Astronom. Soc. **136** (1967), 101; **144** (1969), 189.

D. Lynden-Bell and R. Wood, Monthly Notices Roy. Astronom. Soc. **138** (1968), 495.

D. Lynden-Bell and N. Sanitt, Monthly Notices Roy. Astronom. Soc. **143** (1969), 167.

D. Lynden-Bell and A. J. Kalnajs, Monthly Notices Roy. Astronom. Soc. **157** (1972), 1.

J. R. Ipser, *General relativity and cosmology* 1969, *Course* 47, Italian Phys. Soc. (1971).

J. W.-K. Mark, Astrophys. J. **169** (1971), 356, 455.

L. S. Marochnik and A. A. Suchov, Byull. Inst. Astrofiz. Dushanbe, no. 58, 1971.

A. Toomre and W. Julian, Astrophys. J. **146** (1966), 646.

C.-S. Wu, Phys. Fluids **11** (1968), 316, 545.

ROYAL GREENWICH OBSERVATORY
SUSSEX, ENGLAND

C. Hunter

Fragmentation

1. **Introduction.** It is fairly generally believed that all stars were formed at various stages in the life of the universe through the condensation of very diffuse matter. From this state they evolve, first drawing on their supply of gravitational energy, and then on sources of nuclear energy. It is believed too that star formation is still going on in the spiral arms of our own and other galaxies, where bright O and B stars, which must be young compared with the age of the galaxy, are observed in conjunction with the major concentrations of gas.

The detailed workings of the process by which stars are formed out of diffuse matter are still not well understood. The main agency for causing the condensation of diffuse matter is presumably the force of gravity, though its action can be complicated by magnetic, inertial and thermal effects. Briefly, it is supposed that a large mass of diffuse gas is somehow brought to a state in which it is unstable with respect to its own self-gravitational attraction. In the absence of large magnetic or inertial forces, this state is most likely roughly spherical. The gravitational instability is such that the cloud of diffuse matter starts to collapse in on itself. Now, the total mass that a cloud must have to become gravitationally unstable under likely physical situations is large compared with typical stellar masses. The mass of the unstable cloud depends upon its composition and the state of the matter. Two kinds of clouds have been considered as giving rise to star formation. Hoyle [8] has considered an extragalactic cloud of atomic hydrogen at a temperature of $1.5 \times 10^4 \,^\circ$K and mean density of 10^{-27} gm/cc. For such a cloud, the critical mass for instability as calculated by the methods of §3 is $2.4 \times 10^9 M_\odot$ and so is of galactic magnitude. The critical mass of a hotter cloud is even larger. For a galactic HI cloud, the critical mass is of the rough order of $10^4 M_\odot$. In either situation, the critical mass of the unstable cloud is considerably greater than that of the largest conventional star. If stars are to be formed eventually from the unstable cloud, it must fragment in some way rather than collapse as a single entity. We shall discuss the likely workings of this process of fragmentation, when it arises, and what aids and inhibits it. Before we study these, it is pertinent to remark that this scheme, in which a massive cloud becomes unstable and later fragments into smaller pieces which eventually condense into stars, has the property that groups of stars rather than single stars are formed, in agreement with the observed tendency of stars to belong to associations of various kinds.

2. **Gravitational instability in an infinite uniform medium.** The simplest physical situation in which gravitational instability can occur is that of an infinite uniform medium. Let ρ_0 be its density and c_0^2 the velocity of sound in it. Small perturbations to the medium are governed by the linearized equations

(2.1)
$$\partial \mathbf{u}/\partial t = -\,(c_0^2/\rho_0)\,\mathrm{grad}\,(\delta\rho) + \mathrm{grad}\,\psi,$$
$$\partial\,(\delta\rho)\,/\partial t + \rho_0\,\mathrm{div}\,\mathbf{u} = 0,$$
$$\nabla^2\psi = -\,4\pi G\delta\rho,$$

where $\delta\rho$ denotes the density perturbations and ψ the gravitational potential due to these. (It is supposed here that pressure is a unique function of density so that $c^2 = dp/d\rho$ is well defined.) If we eliminate ψ and $\mathbf{u}$ between these equations, we get

(2.2)
$$\partial^2(\delta\rho)\,/\partial t^2 = c_0^2\,\nabla^2(\delta\rho) + 4\pi G\rho_0\delta\rho.$$

Elementary solutions of this equation of the form $\exp(i\,(\mathbf{k}\cdot\mathbf{x} - \sigma t))$ exist provided

(2.3)
$$\sigma^2 = c_0^2\mathbf{k}^2 - 4\pi G\rho_0.$$

If there were no gravitational term in (2.2), it would be merely the equation for sound waves, and these are not dispersed. The situation is radically altered by the addition of gravitational effects. The value of σ^2 can be negative, and so unstable growing disturbances occur. This happens if

$$c_0^2 \mathbf{k}^2 < 4\pi G\rho_0,$$

that is, if the wavelength

(2.4) $$2\pi/k > c_0(\pi/G\rho_0)^{1/2} = R_J,$$

where we have introduced the symbol R_J, the Jeans' length scale, for the wavelength of marginally unstable waves. Waves of shorter length are stable, while longer ones are unstable. In the present situation, the only forces present are pressure and gravity. The effect of the first is to destroy condensations, while that of the second is to enhance them. When the length scale of the disturbance is sufficiently large, then pressure is unable to stabilize the disturbance against its self-gravity. The instability condition (2.4), though without the right constants, can be obtained by a simple order-of-magnitude argument measuring the relative magnitudes of pressure and self-gravity.

Though the above analysis demonstrates that gravity can cause instability on a sufficiently large scale, the analysis is not directly applicable to cosmogonical problems. The reason for this is that the unperturbed state of an infinite uniform medium does not arise, while any finite mass of gas in equilibrium must be kept in this state by a balance between pressure and gravity, when magnetic forces are neglected. Again, an elementary order-of-magnitude argument shows that this balance implies that the length scale of the gas mass must be of the order of R_J. The infinite medium analysis applied to small subregions of the gas mass gives the result that the regions are stable as only disturbances whose scale does not exceed that of the subregion can properly be considered. The gas mass is therefore stable to disturbances of small scale that tend to break it up. The stability or otherwise of a finite mass of gas in equilibrium must of course be studied ab initio, but our present results suggest that if the mass is unstable, it is likely to be so only to some overall mode of instability such as a radial compression rather than to any subdividing modes.

3. **The virial theorem, and its application to isothermal spheres.**
We shall base our discussion of gravitational instability in a finite
mass of gas on the virial theorem. This is derived by considering an
isolated mass of gas confined within a surface S. The equation of
motion of the gas in a magnetic field and a gravitational field with
potential ψ is

(3.1) $\rho \, D\mathbf{u}/Dt = - \operatorname{grad} p + \rho \operatorname{grad} \psi + (\mu/4\pi) \operatorname{curl} \mathbf{H} \times \mathbf{H}.$

The scalar virial theorem is obtained by taking the scalar product
of Equation (3.1) with the position vector $\mathbf{x}$, and then integrating
over the volume V occupied by the gas. The result, after some
manipulation, is

(3.2)

$$\frac{1}{2} \frac{d^2 I}{dt^2} = 2T + 3(\gamma - 1) \, U + \mathcal{M} + \Omega - 3 p_e V$$

$$+ \frac{\mu}{8\pi} \int [2 x_i H_i H_j - x_j \mathbf{H}^2] dS_j ,$$

where

$$I = \int_V \rho \mathbf{x}^2 dV, \quad \text{half the sum of the principal moments of inertia,}$$

$$T = \frac{1}{2} \int_V \rho \mathbf{u}^2 dV, \quad \text{the kinetic energy,}$$

$$U = \frac{1}{\gamma - 1} \int_V p dV, \quad \text{the internal heat energy with } \gamma \text{ being the ratio of specific heats,}$$

$$\mathcal{M} = \frac{\mu}{8\pi} \int_V \mathbf{H}^2 dV, \quad \text{the magnetic energy,}$$

$$\Omega = -\frac{1}{2} \int \rho \psi dV, \quad \text{the gravitational potential energy,}$$

and p_e is the external pressure, supposed a constant. (For a derivation, see Chandrasekhar [4].)

A necessary condition for an equilibrium state, or even for a
figure which is rotating but has a fixed shape, is that the right-
hand side of Equation (3.2) should vanish. Let us consider first
the simplified case of a mass of gas in which there are no mass
motions and there is no magnetic field. Then the equilibrium requires

$$0 = 3(\gamma - 1)\, U + \Omega - 3p_e\, V.$$

This condition (3.3) is an overall constraint on the detailed structure. We shall apply it to masses of gas which are isothermal. Detailed analyses show that, under the conditions likely to prevail in intergalactic and interstellar clouds, the clouds are optically thin, and can radiate heat in times short compared with possible dynamical time scales, and so maintain themselves at constant temperature. Thus

$$p/\rho = \mathscr{R}T/\overline{\mu}$$

where $\mathscr{R}$ is the universal gas constant, and $\overline{\mu}$ the mean molecular weight.

Strictly spherically symmetric equilibrium configurations are possible, though, so far as I know, there is no proof that these are the only possible equilibrium configurations of finite mass. We shall limit our attention to them. Actually, if the pressure external to an isothermal sphere vanishes, the sphere is of infinite extent and has infinite mass. With the introduction of a nonzero external pressure, and this is realistic as there will always be some background medium surrounding the cloud, the mass and radius of the sphere become finite. The equations for the radial variations of ρ and ψ can not be integrated analytically, though they have been tabulated extensively.

McCrea [15] has shown that it is possible to deduce the essential properties concerning the stability of isothermal spheres with a nonzero external pressure from the virial Equation (3.3) and without using the detailed numerical solution. Let M be the mass of the sphere, and R its radius. Then the gravitational potential energy is given by an expression

$$\Omega = -\, AGM^2/R$$

where A is a dimensionless number. The exact value of A depends on the isothermal sphere solution, and, for a given mass M, it is a function of R. However, A does not vary rapidly with R and so we shall take it to be a constant. With this approximation, and the evaluation

$$(\gamma - 1)\, U = \int_V p\, dV = \frac{\mathscr{R}T}{\overline{\mu}} \int_V \rho\, dV = M\frac{\mathscr{R}T}{\overline{\mu}},$$

our equilibrium condition (3.3) implies

$$p_e = 3M\mathscr{R}T/4\pi R^3\bar{\mu} - AGM^2/4\pi R^4.$$

For a given mass of gas at a fixed temperature, the variation of p_e with R is as shown in Figure 1. There is a maximum external pressure p_{crit} at $R = R_{\text{crit}} = 4AGM\bar{\mu}/9\mathscr{R}T$ for which an equilibrium state is possible. For $p < p_{\text{crit}}$, two values of R are possible. Only the larger value of $R > R_{\text{crit}}$ can represent a stable configuration. For, if $R < R_{\text{crit}}$ and the sphere is contracted slightly, a lower external pressure is needed for the new equilibrium state. The ambient external pressure is greater than this value and so can continue to compress the gas sphere. The sphere is therefore unstable to an overall contraction. (It is easy to see that if $R > R_{\text{crit}}$, the reaction of the sphere after a contraction is to reexpand, and that it is correspondingly stable.) Stable spheres for which $R > R_{\text{crit}}$, and $p_e < p_{\text{crit}}$ can therefore be brought to a state of instability by a process that steadily increases p until $p = p_{\text{crit}}$.

Similarly there is a critical mass that can exist in equilibrium for a given external pressure and temperature (Figure 2). No equilibrium configurations are possible with larger masses, though there are two possible radii for any smaller mass, the larger giving the stable configuration. Once again, we have $R = 4AGM\bar{\mu}/9\mathscr{R}T$ in the critical state. The essential characteristics of these approximate analyses are confirmed by Ebert's [5] rigorous treatment, using the exact solution for spherically symmetric isothermal masses. In support of the approximation of treating A as a constant, it is found that $A = 0.6$ for $R \gg R_{\text{crit}}$ and the gas density is uniform, while the value of A has increased only to 0.73 at the stage the critical condition is reached. An exact analysis shows that the condition

$$R = 0.41\bar{\mu}GM/\mathscr{R}T$$

is satisfied in both the critical states of maximum external pressure for fixed mass and temperature, and maximum mass for fixed external pressure and temperature. Eliminating R through the introduction of the mean density $\bar{\rho}$, this critical condition can be rewritten as

$$M_{\text{crit}} = 1.1 \times 10^{11} \times \left[\frac{1}{\bar{\rho}}\left(\frac{\mathscr{R}T}{\bar{\mu}}\right)^3\right]^{1/2} \text{ grams.}$$

With $\bar{\mu} = 1.4$, $\bar{\rho} = 10$ atoms/c.c. and $T = 125°$K, typical values for an HI cloud, $M_{\text{crit}} = 7.2 \times 10^3 M_\odot$ and $R_{\text{crit}} = 17$ parsecs. These values are reduced at lower temperatures and higher mean densities.

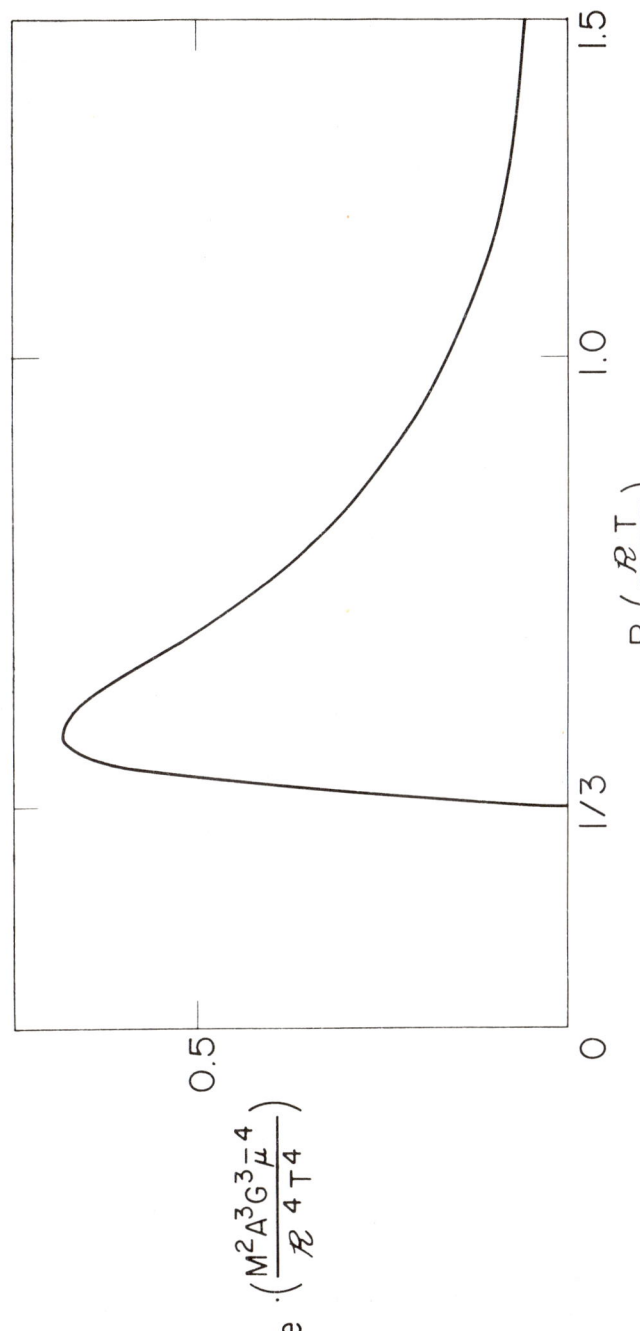

FIGURE 1. Scaled dimensionless pressure versus radius for an isothermal sphere of fixed mass and temperature. In the exact theory, the curve does not cut the R-axis, but spirals to a singularly to the left of the maximum.

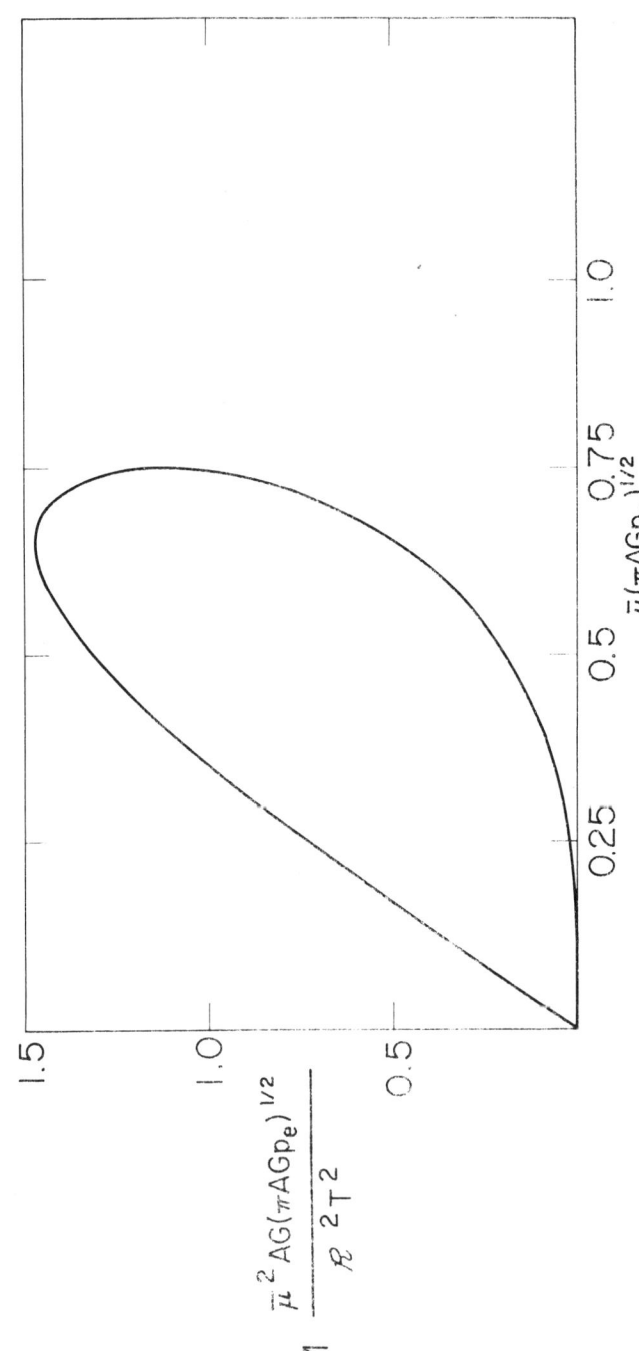

FIGURE 2. Scaled dimensionless mass versus radius for isothermal spheres under given external pressure and at given temperature.

They suggest that the average HI cloud has not reached a state of gravitational instability, though particularly compact or massive ones may well have done so. One possible process by which an unstable cloud can be formed has been discussed quantitatively by Kahn [11]. He considers a cloud of above average density or mass which gains mass through collisions with smaller clouds, and reaches a state of critical stability in a time of the order of 3×10^7 years. The other possibility is to imagine a process by which the external pressure on a cloud, due to adjacent gas and/or collisions with other clouds, is built up.

It is of interest to note that in the diffuse isothermal equilibrium configurations under external pressure with $R \gg R_{\mathrm{crit}}$, $R \ll R_J$ and so Jeans' criterion indicates stability. Only as $R \to R_{\mathrm{crit}}$ does $R \to R_J$. This suggests strongly that the sphere does first become unstable when $R = R_{\mathrm{crit}}$ and not earlier, and that the only unstable mode is an overall spherically symmetric contraction. This result would be made rigorous only by a full stability analysis of isothermal spheres under external pressure, but no such analysis has yet been made. In the present work however, we shall suppose it to be true.

4. **Collapse and fragmentation.** Consider now a spherical cloud that has become gravitationally unstable as described in the previous section, and has started to contract in upon itself. This instability grows because the pressure gradient in the cloud is unable to balance the inward pull of gravity; and it is easily seen that, as the collapse proceeds, this imbalance grows worse. This follows simply from a consideration of the ratio

(4.1)   gravitational forces/pressure forces $\sim GM\bar{\mu}/R\mathscr{R}T$.

Provided the cloud is able to maintain itself in an isothermal state, and the discussions by Hoyle for extragalactic clouds and Mestel and Spitzer [18] for HI clouds have shown that radiation is sufficient to keep pace with the collapse, then the ratio (4.1) increases steadily as R decreases. The collapse of the cloud therefore tends to a free-fall. Some support for this conclusion is given by numerical integrations of the full equations governing the collapse of a gas cloud by McNally [17]. McNally included a detailed equation for the loss of heat in his system, but also constrained the collapse by requiring that the outer boundary of the cloud be held fixed. He found that heat energy could be lost sufficiently rapidly from an interstellar cloud of mass greater than $2.5 \times 10^4 M_\odot$ for the central parts to become approximately freely falling.

The next question to be considered is the dynamical stability of the spherically symmetric collapse of a gas sphere from its critical state. If the symmetric collapse is stable, then of course we would end up with a massive condensed body or protostar, far more massive than any star. If however the symmetric collapse is unstable and the cloud prefers to fragment into a number of subcondensations rather than condense as a single entity, many more and smaller protostars are formed. A crude and inexact argument for the latter eventuality can be based on a straightforward (and strictly incorrect) application of the Jeans' criterion to the collapse. When the sphere just reaches the unstable state, its radius is of the order of the Jeans' wavelength $R_J = (\pi \mathscr{R} T / G \bar{\mu} \rho)^{1/2}$. As the collapse progresses ρ increases as R^{-3}, and so R_J decreases as $R^{3/2}$, faster that is than the radius of the cloud. One would therefore expect the cloud to become unstable to disturbances of successively shorter wavelengths, and thus to fragment. Hoyle [8] discussed fragmentation along these lines, assuming fragments to separate out as soon as they were allowed to by a virial theorem criterion such as (3.5). According to this, successively smaller masses become unstable as $\bar{\rho}$ increases, a result that is essentially equivalent to noting that the ratio R_J / R continually decreases.

Arguments such as the above ignore the dynamics of fragmentation. A proper discussion of any fragmentation must involve an analysis of the symmetric collapse of the isothermal cloud from its critical state. Is this collapse unstable to the formation of subcondensations? Strictly therefore, we should derive the solution of the hydrodynamic equations for the symmetric collapse and then perform a stability analysis on it. Now no analytical solution for this flow is known, and it presumably has to be found, as a function of the radial distance r from the center of the sphere and time t, by numerical integration. Because of its present unavailability and likely complexity, we shall try to make headway by tackling a simpler problem. As we saw earlier, the collapse tends to a free-fall. We shall therefore consider the stability of a free-fall collapse, our analysis thus applying to the original collapse after it has got well under way.

Analytical solutions describing a free-fall collapse are simpler to obtain. Using a Lagrangian description of the flow,

(4.2) $$\partial^2 r / \partial t^2 = -\, Gm(a) / r^2$$

where $m(a)$ is the mass within the sphere of radius r. This mass is unchanged for any material sphere during the flow if shells of fluid do not pass through each other, and can therefore be regarded as a function of the radius a of the sphere at some reference time. We take a as a Lagrangian variable so that $r = r(a, t)$, and the fluid density ρ is given by

(4.3) $$4\pi \rho r^2 \, \partial r / \partial a = m'(a).$$

The equation of motion (4.2) can be integrated once to give

(4.4) $$\frac{1}{2} \left(\frac{\partial r}{\partial t} \right)^2 = \frac{Gm(a)}{r} + f(a)$$

where f is a function of integration, which can be fixed only after boundary conditions have been applied. It is not clear just what boundary conditions should be applied if we are to regard our flow as a late stage in the development of the collapse from a critically stable state, so we shall for simplicity choose $f(a) = 0$. This means that fluid particles are at rest when they are infinitely far away from the origin. Then we can integrate (4.4) again to

(4.5) $$r = (9Gm(a)/2)^{1/3} (t_0(a) - t)^{2/3},$$

$t_0(a)$ being another function of integration, and representing the time at which a particular fluid particle reaches the center $r = 0$. (Actually, not all solutions (4.5) are physically acceptable. Any in which infinite density gradients occur, and shells of particles pass through one another, must locally violate the original assumption of negligible pressure forces.)

A particularly simple flow is given by taking $t_0 = 0$, so that all the matter collapses to a point mass at the instant $t = 0$. It is readily seen that

(4.6) $$\rho = 1/6\pi G t^2,$$

and so the density is spatially uniform at all times. (There are actually a whole family of solutions for which the density is spatially uniform, but they all give a density which tends to the form (4.6) as $\rho \to \infty$. As will be seen later, it is in this late stage of the collapse when ρ grows large that the main growth of the instability occurs.) The radius of the cloud varies as

(4.7) $$R = (9GM/2)^{1/3} (- t)^{2/3}$$

and the Eulerian velocity

(4.8) $$u = 2r/3t.$$

It is most convenient to go over to an Eulerian representation of the flow field for the linearized stability analysis, which is carried out in the usual manner. It is also convenient to change the independent variables from the position vector $\mathbf{x}$ and t, to $\mathbf{x}^1 = (- t)^{-2/3}\mathbf{x}$ and t. The scaled position vector $\mathbf{x}^1$ therefore contracts along with the collapse, and $\mathbf{x}^1 = $ constant describes a particle path in the unperturbed flow. In terms of these new variables, the density perturbation $\delta\rho$ is readily seen to satisfy the equation

(4.9) $$\left(\frac{\partial^2}{\partial t^2} + \frac{16}{3t}\frac{\partial}{\partial t} + \frac{4}{t^2}\right)\delta\rho = 0,$$

so that

(4.10) $$\delta\rho = (- t)^{-3}g_1(\mathbf{x}^1) + (- t)^{-4/3}g_2(\mathbf{x}^1).$$

(For details of the full stability analysis, see Hunter [9] and [10].) Here g_1 and g_2 are arbitrary functions of integration, and they are made precise only when particular forms of the perturbations at some initial instant are specified. The power law behavior in time rather than the more familiar exponential growth is due to the fact that the basic flow is time varying.

It is immediately clear that the solution (4.10) shows the spherically symmetric collapse to be unstable. Although the unperturbed density ρ becomes large as $- t \to 0$, $\delta\rho$ grows faster and

(4.11) $$\delta\rho/\rho \sim \rho^{1/2}.$$

The perturbations in the density field grow relative to the uniformly contracting background, and cause the break-up of the collapse to a single point. The predominant form of the perturbations is given by the function g_1, and we have incipient subcondensations starting to grow around points at which g_1 has maxima. Note that one gets the perturbation growth law (4.11) by comparing a density

$$\rho + \delta\rho = 1/6\pi G(t + \epsilon)^2 \qquad (|\epsilon| \ll 1)$$

with (4.6), that is an exactly similar collapse but with a slightly different instant of concentration. Such a collapsing flow is of course included in our perturbation analysis, which includes all flows that differ slightly from the original collapse. It is a very special kind of perturbation as g_1 is a constant and the property of collapse to a single point is preserved. This is not the case with more general forms of the function g_1.

The detailed stability analysis shows that the cause of the instability is simply the self-gravitation of the incipient fragments which acts to accentuate their growth. In fact, it is possible to continue the stability analysis through to second-order terms, and these enhance the growth of subcondensations compared with the results of the linearized theory. An illustration of this is given by the particular solution

$$(4.12) \quad \rho = \frac{1}{6\pi G t^2}\left[1 + \frac{C}{(-t)}\sin\frac{2\pi z^1}{L} - \frac{C^2}{(-t)^2}\cos\frac{4\pi z^1}{L}\cdots \right]$$

where C and L are constants, for which the first-order perturbation has a one-dimensional sinusoidal form. The second-order term is positive near both the maxima and minima of the sine function and negative in between, thus accentuating the peaks and filling in the dips in the original sinusoid.

Expressions for the kinetic energy T and gravitational potential energy Ω can also be calculated through to the second order. The energy equation for the freely falling flow is simply

$$T + \Omega = 0.$$

Denoting by subscripts the terms of different order, then it can be shown that $T_2 = -\Omega_2$, the second-order contribution to the kinetic energy and release of gravitational energy, is positive whatever the forms of the perturbations. This indicates that a physical reason for the instability is that the flow is able to release gravitational energy faster by forming subcondensations, rather than by collapsing symmetrically as a whole, and prefers to do this. Considerations such as these suggest that other free-falling collapses besides the particular one we have analyzed will also suffer the same kind of fragmenting instability.

Formula (4.10) shows that disturbances of any wavelength are treated identically. This is principally due to the approximation of free-fall and the neglect of pressure forces. The shorter the wavelength of a disturbance, the less good this approximation will be. However, as follows from a consideration of the ratio of gravitational to pressure forces, this approximation improves continually as the flow progresses, provided of course that, as before, it remains isothermal. Thus, if we had the exact solution for the collapse of an isothermal sphere from its critical state, we should expect that initially the sphere is unstable only to its overall contraction, but that modes of progressively shorter wavelength become successively unstable as the collapse progresses. At later times, we expect the disturbances of larger wavelength to predominate as they have had longer to grow.

An important modification to the flow we have been considering can occur when gas densities increase sufficiently for the opacity of the gas to grow significantly, and stop the transfer of heat needed to maintain isothermal conditions. This change of conditions will in general occur only after densities have increased by many orders of magnitude so that initially small fluctuations will have had a good chance to grow and initiate fragmentation. The significance of the increase of opacity is that it becomes possible for the pressure forces to start growing in importance relative to the gravitational forces, and eventually to become dynamically significant. When they do so, they must ultimately halt the collapse of fragments, though this halting of the collapse would almost certainly be accompanied by violent radial pulsations of the fragments, rather than being an abrupt stoppage. These pulsations will be damped by radiation, and shock waves may also form and transform kinetic energy into heat energy.

If the behavior of the gas can be described adequately by a simple polytropic law $p = k\rho^\gamma$ with k and γ constants, then the ratio of pressure to gravitational forces as given by (4.1) increases or decreases with collapse according as γ is greater than or less than $4/3$. Thus in the extreme case of a gas which is completely opaque so that there is no heat transfer and also no communicable internal energy $\gamma = 5/3$ and the collapse is brought to a halt. Actually the thermodynamics when opacity increases substantially are likely to give rise to a more complicated situation than can be described by a simple polytropic law.

Hoyle [8] gives an order-of-magnitude estimate of the stage at which fragmentation ends and bound protostars are formed, but this is based on the particular hierarchial model of fragmentation he adopts. In his model, successive generations of fragments are formed as soon as they are allowed to do so by a virial theorem criterion. He neglects the dynamics of the process, and moreover assumes that at each stage of fragmentation, the gravitational energy released is all radiated away and does not go at all into kinetic energy of motion. In fact he argues that if the gravitational energy released did go into mass motions, these must be adequate to re-expand the cloud to nearly its original volume and so stop fragmentation. This fragmentation scheme is thus very different from that outlined in the present work, and Hoyle's calculation of the end point of the fragmentation process is not relevant to it. His calculation supposes that fragmentation ceases when the temperature in the gas has not risen appreciably but conditions are suitable for a Kelvin-Helmholtz contraction phase to begin. The protostars then have radii of the order of 10^{23} cm and mean densities of 10^{-8} gm/cm^3, so there is still a considerable Kelvin-Helmholtz contraction to be gone through before a stellar state is achieved. In the present scheme, when considerable kinetic energy of motion is generated, the process of halting fragmentation and forming bound protostars will be different. It follows from simple energy arguments (see Mestel [19]) that if a gas sphere of initial radius R_0 is allowed to collapse freely and symmetrically to a radius R_1 ($\ll R_0$), and then we suddenly change to an adiabatic gas law with $\gamma > 4/3$, the sphere continues to collapse to a radius $R_2 \ll R_1$ and then undergoes oscillations of large amplitude which may be damped by various processes. In the compressive stages of these, high temperatures must be built up, and complications such as ionization of the hydrogen could result. An analysis of what happens is lacking, but it does seem likely that considerably higher densities and temperatures than those envisaged by Hoyle will be attained before the formation and growth of fragments is halted, and that the Kelvin-Helmholtz contraction phase will be much less extensive.

It is interesting to note in this context that Gaustad [6], after a careful study of physical processes in the temperature range 50° K to 3000° K and density range 10^7 to 10^{17} a.m.u., concluded that no slow Kelvin-Helmholtz contraction state was possible in these ranges except perhaps for protostars of a tenth of a solar mass or less. Cameron [3] in his study of the formation of the solar system from an HI cloud followed the fragmentation of an unstable cloud (initial density 10^3H atoms/cc) assuming this fragmentation to cease at the stage calculated by Mestel and Spitzer (which was also based on Hoyle's fragmentation scheme) when the central temperature of a fragment was about $1000°$K. Taking the further development to be a Kelvin-Helmholtz contraction, Cameron found that a second collapse ensued when temperatures exceeded $1800°$K and vibrational levels of H_2 were excited and considerable dissociation occurred, reducing the effective value of γ below $4/3$. This collapse ended with the fragment of one solar mass having a radius of roughly 50 solar radii, with temperatures of the order of $10^{4°}$K and all the hydrogen and helium ionized, the degree of condensation in this second collapse being of the same order as that achieved in the first collapse. It is clear therefore that the cut-off of the fragmentation process requires further study, and that the protostars finally formed will probably be more nearly stellar than was suggested in the earlier work of Hoyle and of Mestel and Spitzer. The dynamics can not safely be neglected in such a study. Indeed, what is wanted is a discussion of the chaotic turbulent state of motion initiated by the break-up of the original orderly collapse, and its outcome. This turbulent state will involve supersonic velocities, large density contrasts, and complicated thermodynamic processes. Although the stability analysis relates the perturbation quantities at all times to their initial values at some earlier reference time, it is more likely that, once the turbulence has been triggered, the final forms of the fragments result from the basic character of this turbulence rather than from any initial conditions in an early stage of the collapse. Needless to say, this turbulence problem is likely to be difficult.

At the stage at which nonuniformities in the density field grow in our linearized stability analysis, perturbations in the velocity field also grow and, when $\delta\rho/\rho \to O(1)$, then nonradial velocities tend to become equal in magnitude to the symmetrical collapse velocities. The motion of the fragments is therefore breaking away from the collapse to a point. Assuming that a good proportion of the fragments are not broken up in collisions, nonradial velocities will also be created through encounters. Although it is not clear what the final state of the system will be, it seems likely that the system will achieve its minimum radius roughly when the inward collapse breaks up into fragments. Because of the large amount of gravitational energy released which goes into the kinetic energy of the fragments, we must expect that the final radius of the resulting stellar system, which we imagine as forming eventually, will be considerably larger than the minimum radius when the fragments first grow, and the system may not even be bound. The final radius cannot be calculated until we have a better idea of the energetics, how much gravitational energy is finally released before fragmentation ceases, how much energy is dissipated in fragment formation and so on.

In the development outlined above, it would seem that the greatest likelihood of collisions occurs just after the fragments have first formed. At this stage, their own inward collapse is faster than that of the overall collapse so that many direct collisions could be avoided because of the rapid decrease of cross-sectional area of the fragments. This phase too requires a more detailed investigation. An analysis by T. T. Arny [2] suggests that a high proportion of the fragments may survive destruction by collisions.

5. **Rotation and the angular momentum problem.** In the work of the previous section, we supposed the critically stable isothermal cloud from which the collapse started to be at rest. The matter from which we hope to construct our stars may be in a state of rotation; interstellar gas which partakes of the general galactic rotation certainly is. Now, although there is no analysis of a mass of rotating isothermal gas reaching a critical state similar to that for a nonrotating mass, it is evident that a small amount of rotation is not going to make any significant difference and that a similar critical state can still be reached. Here the rotation would be small in the sense that the kinetic energy T of the rotational motion would be very much less than $(\gamma - 1) U$ and $- \Omega$. If the critically stable HI cloud for which figures were quoted in §3 were supposed to be rotating uniformly with angular velocity equal to half the local vorticity of the galactic motion (Oort's constant B), then its kinetic energy of rotation is of the order of 10^{-3} of its gravitational energy. Although such a slowly rotating mass will be almost spherical and its initial collapse will be almost spherically symmetric, this near-sphericity will not prevail indefinitely. For one thing, the gravitational field of a nonspherical body generally tends to increase this nonsphericity. As an example, freely falling spheroids of uniform density collapse through a series of spheroidal states with steadily increasing eccentricity (Lynden-Bell [14]; Lin, Mestel and Shu [12]).

(The increase of eccentricity for an initially almost spherical spheroid can be regarded as a special instance of the instability of the spherical collapse discussed in the previous section.) As well as this purely gravitational effect, there is also the inertial effect of the rotation. Consider a cloud with uniform angular velocity ω about an axis through the center of the cloud. If the cloud collapses isotropically, then the centrifugal force $R\omega^2$, as a result of the conservation of angular momentum ωR^2, increases as R^{-3} which is faster than the growth of gravitational force as R^{-2}, so that centrifugal effects, though initially small, must inevitably become significant. The collapse cannot remain spherically symmetric, but will tend to a disk-like configuration, since the rotation will halt the collapse perpendicular to the axis of rotation but will not affect the collapse parallel to the axis.

Actually, when the full perturbation analysis of the spherical collapse as described in the previous section is performed, a rotational component of the velocity field is present in general, and it is included in the analysis. The relative significance of this rotation grows as R^{-1}, as in the above argument, that is as $\rho^{1/3}$. This rate of growth is slower than the $\rho^{1/2}$ growth of density fluctuations, as given by Equation (4.11). Thus the rotation does not hinder the start of fragmentation, at least according to the linear theory, if the initial nonuniformities of the critical state are of a general form. It may do so, however, if they are predominantly rotational, and the spherical collapse to a point is significantly modified by rotation before fragments begin to separate out.

The growth of particle density in the fluid can be studied in more general terms by means of an exact equation that can be deduced from the full hydrodynamic equations. If we take the divergence of the equation of motion (3.1) with the magnetic force term omitted to evaluate $D(\text{div }\mathbf{u})/Dt$, and then use the continuity equation in the form

$$(5.1) \qquad\qquad \text{div }\mathbf{u} = \frac{-1}{\rho}\frac{D\rho}{Dt},$$

then, after some rearranging, one gets the equation

$$(5.2) \quad \frac{D^2}{Dt^2}(\log\rho) = -\nabla^2\psi + \text{div}\left(\frac{1}{\rho}\,\text{grad}\,p\right) + e_{ij}e_{ij} - \frac{1}{2}(\text{curl }\mathbf{u})^2$$

where e_{ij} is the rate-of-strain tensor defined as

(5.3)
$$e_{ij} = \frac{1}{2} \left(\frac{\partial u_i}{\partial x_j} + \frac{\partial u_j}{\partial x_i} \right) .$$

We can also use the gravitational field equation

(5.4)
$$\nabla^2 \psi = -4\pi G \rho$$

in Equation (5.2). We now have an equation describing particle density variations, which shows that the growth of ρ is accelerating if the right-hand side term is positive. Now the first term on this right-hand side is $4\pi G \rho$ and is essentially positive. It exhibits the effect of the self-gravity of the fluid element in acting to enhance itself. The second term represents the disruptive effect of pressure which acts to smooth out any condensation, at least if $dp/d\rho > 0$. The third term involving the straining component of the velocity field is essentially positive, and so does not hinder the growth of density as has sometimes been suggested. Only the rotational part of the velocity field acts to decrease $\log \rho$. The only effects which can stop condensations, in the absence of magnetic forces, are pressure and vorticity. Now, so long as the gas flow is barotropic, that is the pressure is a function of density only, as is the case with both isothermal and adiabatic flow, the growth of vorticity is governed by Kelvin's circulation theorem. This means that the circulation around a material fluid circuit is constant. Vorticity is not created or destroyed, and what is present initially can be enhanced only by the compression of fluid elements which decreases the area of material circuits, and thereby increases the local vorticity.

Let us now consider the simple case of an isothermal cloud in which all the elements are rotating about a single axis. Suppose too that the rotation is sufficiently small initially so that the cloud passes through an almost spherical critical state essentially as before, but suppose that conditions are such that fragmentation does not occur before the rotation has grown significantly. The effect of the rotation is to hinder collapse perpendicular to its axis, and so cause a flattening of the cloud along the axis of rotation. If we suppose that all the kinetic energy of motion parallel to the axis of rotation is destroyed, then it is possible for the cloud to become a very thin disk in steady rotational motion. In such a disk, the balance of forces perpendicular to the axis of rotation is principally between the gravitational and centrifugal forces, while that parallel to the axis of rotation is between gravity and the much weaker pressure force. In fact, a disk-like form is the only possible steady state configuration that could be achieved from the initial conditions we assumed. The cloud must contract significantly towards the axis of rotation until the initially small ratio of rotational to gravitational energy which grows as R^{-1} has become of order unity. During this collapse, the ratio of thermal to gravitational energy has decreased like R, and so has become small.

Analytical expressions describing the disk equilibrium can be obtained. Assuming the gas to continue in an isothermal state, and taking the z-axis as that of rotation, the balance of forces in the z-direction requires

$$(5.5) \qquad \frac{\partial \psi}{\partial z} = \frac{\mathscr{R}T}{\bar{\mu}\rho} \frac{\partial \rho}{\partial z}.$$

The disk structure implies $\partial/\partial z \gg \partial/\partial x, \partial/\partial y$, and so the gravitational field equation is approximately

$$(5.6) \qquad \partial^2 \psi / \partial z^2 = -4\pi G\rho.$$

These equations can be solved to give the structure of the disk in the z-direction as

$$(5.7) \qquad \rho = \frac{\bar{\mu}\pi G\sigma^2}{2\mathscr{R}T} \operatorname{sech}^2 \left\{ \frac{\bar{\mu}\pi Gz\sigma}{\mathscr{R}T} \right\}.$$

Here $\sigma = \int_{-\infty}^{\infty} \rho\, dz$ is a function of integration, and is the surface density obtained when the disk thickness tends to zero. An explicit formula for σ can be given only after we have balanced centrifugal and gravitational forces in the plane of the disk. In this plane, the first and fourth terms on the right-hand side of (5.2) are in approximate balance. Without solving in detail, it is apparent that

$$\text{disk thickness} = O\left\{ \frac{\mathscr{R}T}{\bar{\mu}\pi G\sigma} \right\} = O\left[R \times \left(\frac{\text{velocity of sound}}{\text{rotational velocity}} \right)^2 \right]$$

$$(5.8)$$

$$= O\left[R \times \left(\frac{\text{thermal energy}}{\text{rotational energy}} \right) \right],$$

where R is now the disk radius. The disk we have imagined forming is geometrically thin, and this thinness ratio is of the order of the degree of contraction of the cloud perpendicular to the axis of rotation from its original spherical to its final disk form.

It is interesting to note that a one-dimensional isothermal equilibrium as described by Equation (5.7) is always possible.

There is no one-dimensional analogue of the critical state for spherical isothermal configurations, and in fact a detailed stability analysis shows the configuration to be stable to disturbances with variations in the z-direction only (Goldreich and Lynden-Bell [7]). However, thin disks are unstable to disturbances in which there is motion predominantly in the plane of the disk. This instability is exhibited by the idealized case of an infinitesimally thin disk, another zero temperature approximation like that of free-fall.

A dimensional analysis of the stability of an infinitesimally thin disk has been given by Toomre [23]. Consider a local condensation in the disk of length scale L involving changes of relative magnitude ϵ. The mass involved in the condensation is $O(\sigma L^2)$, particles are displaced $O(\epsilon L)$, so that the extra gravitational forces produced are $O(G \cdot \sigma L^2 \cdot \epsilon L / L^3) = O(\sigma G \epsilon)$. In the zero temperature approximation with pressure neglected, the only possible forces to prevent a condensation from growing are inertial forces. These are $O(\omega \times \text{perturbed velocity})$, that is $O(\epsilon L \omega^2)$. We therefore have instability if

(5.9) $$\sigma G \gg L\omega^2, \qquad L \ll \sigma G / \omega^2 = O(R).$$

This leads us to expect the disk to be unstable to all disturbances of wavelengths considerably less than the radius of the disk, a result that is confirmed by detailed stability analyses. These analyses show that, although various disks possess some stable modes of oscillation, all of long wavelength, they also have infinitely many unstable modes of short wavelength, and in fact the rate of growth of unstable modes increases with decreasing wavelength. Thus, although rotational forces can achieve an overall balance in the disk state, they are not able to stop subcondensations forming as a result of the gravitational effects of local density fluctuations.

The approximation of treating the disk as infinitesimally thin is valid only for disturbances of wavelength long compared with the disk thickness. This thickness is of the order of the Jeans' length scale (2.4) based on, say, the central density of the mean density across the disk. On length scales of the order of the disk thickness, the stabilizing force of pressure will come into play. The maximum growth rate will presumably occur for waves of some intermediate length between the disk thickness and its radius.

These will cause the surface density to increase locally, and so some further flattening must ensue, the disk thickness being inversely proportional to the integrated surface density. If the condensations are still disk-like, further break-up will follow, and stabilization can be achieved only when the pressure becomes a major force in all directions in each subcondensation. This would be the case if the original disk broke up into fragments through disturbances of length roughly equal to the disk thickness in the plane of the disk. If we apply this picture to the case of the particular HI cloud mentioned above for which the kinetic energy of rotation was initially of the order of 10^{-3} of its gravitational energy, and which thus condenses to a disk with thickness to radius ratio of 10^{-3}, we get of the order of 10^6 subcondensations each with mass of the order of $10^{-2} M_\odot$. The mean density of these is 10^{12} times the initial density. Even if the break-up into fragments stops here, the net result is bodies which are much less massive than stars. This, as Mestel [19] points out, is one aspect of the "angular momentum problem." In its more usual form, the "angular momentum problem" is presented by noting that if the sun were formed from material of the present interstellar cloud density by isotropic compression with angular momentum conserved, then that material would have had to have an angular velocity of only 5.8×10^{-5}km/sec kpc^{-1}, far less than any reasonable value under present day conditions at least. (Oort's constant B is of the order of -8km/sec kpc^{-1}.) This initial angular velocity must be raised by a factor of the order of 100 when the same calculation is made for O and B stars (Allen [1], p. 204), but it is still very small. One possible resolution, discussed in the next section, is to invoke magnetic forces to transfer angular momentum from a condensation to surrounding gas, and so lessen the rotational braking. It should also be pointed out that, although, for simplicity, the rotation is normally assumed to be uniform, it need not be, and the vorticity will in general be distributed nonuniformly. As Equation (5.2) shows, the formation of condensations is favored in regions of below average vorticity. We are not concerned with any particular part of the fluid forming a subcondensation, all we require is that subcondensations form somewhere. A similar if more discrete argument was given by McCrea [16], who pictured a gas cloud as consisting of a large number of cloudlets in supersonic motion. He envisaged condensations as growing first through chance collisions and then gravitational attraction as the condensations become larger. Cloudlets with large angular momentum one about the other do not combine; only those with small relative angular momentum do. However, with the large discrepancies in angular velocity noted above, it is doubtful that the angular momentum problem can be altogether avoided by supposing all condensations to arise from favored regions of low vorticity. It seems that we must look to magnetic fields to help resolve this difficulty.

6. **Some effects of a magnetic field.** The presence of a galactic field adds another complicating feature to the problem of star formation from interstellar clouds. The additional magnetic terms in the virial Equation (3.2) must be retained and, before a critical state of marginal stability of the kind we considered previously can be attained, the gravitational energy must be sufficient to overcome the sum of the thermal and kinetic energies and magnetic terms. The relative importance of these magnetic terms depends of course on the magnitude of the magnetic field. If the critically stable HI cloud for which figures were quoted earlier contains a uniform field of 10^{-6} gauss, its magnetic energy is 2.4×10^{46} ergs, while its gravitational energy is 1.9×10^{47} ergs. The magnetic energy of a uniform field of 10^{-5} gauss is 100 times larger and so would dominate the gravitational energy, and completely alter our previous picture of gravitational collapse.

Some idea of the action of a magnetic field can be gained by considering simplified situations in which thermal effects are neglected, and the effects of gravity and magnetic field only are considered. Consider for simplicity a uniform sphere of radius R_0, density ρ_0, at rest and containing a uniform magnetic field H_0. Suppose too that the magnetic field in the sphere joins onto a current-free exterior field which tends to zero at infinity. Such a sphere is not in strict equilibrium, as the magnetic force term in the equation of motion (3.1) vanishes everywhere except at the surface of the sphere, where there is a surface current and it is infinite. However, if we neglect the detailed balance of forces, and consider only total integrated effects via the virial theorem, we have

$$\frac{1}{2}\frac{d^2I}{dt^2} = \frac{-3GM^2}{5R_0} + \frac{H_0^2 R_0^3}{4}.$$

(See Strittmatter [22].) Note that the magnetic surface integral adds a nonvanishing term to the right-hand side equal to half the magnetic energy of the sphere. For an overall contraction to be possible, it is necessary that

$$M > M_{\text{crit}} = \frac{H_0 R_0^2}{(2.4G)^{1/2}} = \frac{15 H_0^3}{128 \pi^2 \rho_0^2 G} \left(\frac{5}{3G}\right)^{1/2}$$

With $H_0 = 10^{-6}$ gauss, $\rho_0 = 10$ atoms/cc as before, we get $M_{\text{crit}} = 800 M_\odot$, but it is $8 \times 10^5 M_\odot$ if $H_0 = 10^{-5}$ gauss, which again shows the dominant part any field as high as 10^{-5} gauss must play.

Consider next the result of an isotropic contraction of the sphere. We shall suppose that, because of the high conductivity of interstellar matter, the field remains frozen into the material. Then, when the radius of the sphere is R, the magnetic field H is given by $H = H_0(R_0/R)^2$, and the right-hand side of the virial Equation (3.2) is

$$\frac{R_0}{R} \left(\frac{-3GM^2}{5R_0} + H_0^2 R_0^3\right)$$

This is simply a factor R_0/R times its value at radius R_0, and so, as noted by Mestel and Spitzer, the critical mass is unaltered by the isotropic contraction. If the original sphere is critically unstable, it remains so throughout the isotropic collapse. This situation differs from the collapse of a nonmagnetic isothermal cloud from a critically stable state, in which the critical mass according to the simple virial theorem criterion steadily decreased. The rough argument based on this fact was that fragmentation occurred during the collapse, a prediction that was supported by the dynamical stability analysis. The analogous argument applied to the case of the magnetic cloud predicts that no fragmentation occurs, though this result is not rigorously established.

The collapse of magnetic clouds with frozen-in fields has been examined in more detail by Strittmatter but still using the virial theorem criterion. The previous assumption of isotropic collapse through spherical states was an oversimplification, and we would expect preferential collapse along the field lines. Strittmatter considers collapse through a sequence of uniform oblate spheroidal states with the magnetic field remaining uniform and frozen-in and shows that, even if the spheroid collapses to a disk, the critical mass is reduced only by a factor of a half or so. The possibilities for fragmentation during the collapse of a cloud with a large magnetic field would thus seem to be drastically curtailed.

These difficulties concerning fragmentation can be avoided if the condition that the magnetic field is frozen into the gas is released. If some agency can be found for getting rid of the magnetic field in a cloud, the problem is simply reduced to the nonmagnetic case. The possibility of this was first discussed by Mestel and Spitzer [18] who showed that a mutual drift between the neutral gas and the much smaller number of charged particles could be set up which allowed the magnetic field to slip through the gas with the charged particles. This was able to happen when the plasma density was reduced through the obscuration of galactic starlight by dust grains.

Although, owing to revisions in certain physical parameters, this process is likely to be far less effective than Mestel and Spitzer calculated, Spitzer [21] has concluded that reasonable separation of magnetic field and matter is possible during the later stages of collapse. As Spitzer [20] has remarked elsewhere, the magnetic field must eventually get out, for even with the low field of 10^{-6} gauss, interstellar matter compressed isotropically to solar density would be such that the magnetic energy density would exceed the present average material energy density in the solar interior by several orders of magnitude.

Another relevant and important property of a magnetic field is its ability to transfer angular momentum (Lüst and Schlüter [13]). This can be exhibited in a special case by considering the equations of motion (3.1). We shall consider for simplicity an axially symmetric flow with the z-axis as that of symmetry. Then, using cylindrical polar coordinates, $\partial/\partial\theta = 0$, and the θ equation of motion can be written

$$(6.4) \quad \frac{D}{Dt}(rv) = \frac{\mu}{4\pi}\left[H_r\frac{\partial}{\partial r}(rH_\theta) + H_z\frac{\partial}{\partial z}(rH_\theta) \right] = \frac{\mu}{4\pi}\,\mathbf{H}\cdot\mathrm{grad}(rH_\theta).$$

The orbital angular momentum rv of each particle is conserved when there is no magnetic field since the right-hand side then vanishes, but not otherwise. If the magnetic field is frozen-in, then during an inward collapse of a region of the gas, the field lines will get twisted and will exert a counteracting torque, which will tend to equalize any difference in angular velocity between the condensation and its background, and so reduce the growth of angular velocity that would otherwise occur. If this mechanism is to be useful in helping solve the angular momentum problem for star formation from HI clouds, then one does not want the magnetic field to be able to slip too easily out of the gas as originally suggested by Mestel and Spitzer, so that the more recent revisions are helpful in this respect. Although the transfer of angular momentum will be curtailed when the magnetic field lines eventually slip through the gas, there is also the possibility of a local bunching of oppositely directed field lines after a substantial collapse, leading to a local breakdown of the freezing-in condition, and the field lines snapping off. This process, in detaching the magnetic links of the condensation with its background, will also tend to put a stop to the transfer of angular momentum. This and other aspects of condensations in rotating magnetic media are discussed extensively by Mestel [19].

References

1. C. W. Allen, *Astrophysical quantities*, University of London, Athlone Press, 1963.

2. T. T. Arny, Ph. D. Thesis, University of Arizona, 1965.

3. A. G. W. Cameron, Icarus, 1 (1962), 13.

4. S. Chandrasekhar, *Hydrodynamic and hydromagnetic stability*, Clarendon Press, Oxford, 1962.

5. R. Ebert, Z. Astrophys. 37 (1955), 217.

6. J. E. Gaustad, Astrophys. J., 138 (1963), 1050.

7. P. Goldreich and D. Lynden-Bell, Monthly Notices Roy. Astronom. Soc. 130 (1965), 97.

8. F. Hoyle, Astrophys. J., 118 (1953), 513.

9. C. Hunter, Astrophys. J., 135 (1962), 594.

10. _____, Astrophys. J., 139 (1964), 570.

11. F. D. Kahn, *Die Entstehung von Sternen*, Springer-Verlag, Berlin, 1960.

12. C. C. Lin, L. Mestel and F. H. Shu, Astrophys. J., 142 (1965), 1431.

13. R. Lüst and A. Schlüter, Z. Astrophys. 38 (1955), 190.

14. D. Lynden-Bell, Proc. Cambridge Philos. Soc. 58 (1962), 709.

15. W. H. McCrea, Monthly Notices Roy. Astronom. Soc. 117 (1957), 562.

16. _____, Proc. Roy. Soc. A 260 (1961), 152.

17. D. McNally, Astrophys. J. 140 (1964), 1088.

18. L. Mestel and L. Spitzer, Monthly Notices Roy. Astronom. Soc. 116 (1956), 503.

19. L. Mestel, Quart. J. Roy. Astronom. Soc. 6 (1965), 161, 265.

20. L. Spitzer, *Interstellar matter in galaxies*, L. Woltjer, ed., Benjamin, New York, 1962.

21. _____, *Origin of the solar system*, R. Jastrow, ed., Academic Press, New York, 1963.

22. P. A. Strittmatter, Monthly Notices Roy. Astronom. Soc. 132 (1966), 359.

23. A. Toomre, Astrophys. J. 139 (1964), 1217.

MASSACHUSETTS INSTITUTE OF TECHNOLOGY
CAMBRIDGE, MASSACHUSETTS

E. Margaret Burbidge

Radiogalaxies

I. **Brief historical account.** The first discrete radio source, called Cygnus A, was discovered in 1946 by Hey, Parsons, and Phillips. In 1948, Bolton discovered the further strong sources Virgo A, Taurus A, Hercules A, and Centaurus A, and Ryle and Smith discovered Cassiopeia A, the strongest source in the sky. In 1949 Bolton and Stanley identified Taurus A with the Crab Nebula, the remnant of a supernova explosion, and Bolton, Stanley, and Slee made the first extra-galactic identification of Virgo A with M 87 and Centaurus A with NGC 5128, which are both bright, comparatively nearby galaxies. The very accurate radio position (Smith [25]) for Cygnus A enabled Baade and Minkowski [3] to identify this source with a curious apparently double galaxy in a rich cluster. The visual magnitude, corrected for rather heavy obscuration in our own Galaxy, was 14.3 (the observed apparent magnitude was about 16) and the redshift was found to correspond to a velocity of 16,500 km/sec. Yet despite this rather large distance and the fairly faint optical luminosity of the object, it was the second strongest radio source in the sky.

As a consequence of this identification, two facts were immediately realized: (1) the radio powers emitted at the sources could sometimes be very large; (2) some of the less bright radio sources might be very distant. For Cygnus A, the total radio power output at the source is 10^{45} erg/sec. Such a source, placed at a very great distance, could still be detected in the radio wavelength region even after its optical brightness had become too faint for its detection by even the largest optical telescopes.

II. **Radio structure and time scales for the radio emission.** Many of the radio sources have been found to be double, with the optical galaxy lying at the centroid of the radio distribution. This was first discovered in the case of Cygnus A by Jennison and Das Gupta [15] by interferometry. By observing sources with two radio antennas separated by a variable spacing, and by analyzing the interference pattern obtained, it is possible to deduce the radio brightness distribution. Maltby and Moffett [17] have concluded that 73% of those radio sources which can be resolved, i.e., those which do not appear as point sources, are actually double or multiple with separations going up as high as 250 kpc. Simple Gaussian distributions account for 17% of the resolved sources, and distributions consisting of a sharp core with a surrounding halo account for the remaining 10%.

These studies of the radio brightness distribution have shown, therefore, that the radio emission often comes from large volumes outside the galaxies; this is an important point in consideration of theories of the radio emission and their consequences, for example, with regard to the possibility of a universal intergalactic cosmic ray flux.

Interferometer measures have recently been made at the Jodrell Bank radio telescope, by Palmer and his colleagues, with spacings going up to 200,000 λ (i.e., separations of 100-200 miles between the two antennas). It was found that 5% of the sources studied had angular diameters less than 1 second of arc. Recently, the method of studying radio sources by lunar occultation has provided valuable data on the radio structure of some sources. This has been done in particular by Hazard and colleagues in Australia. A further method which is useful for picking out sources of very small angular diameter is to look for those sources which scintillate in the interplanetary medium (Hewish, Scott and Wills [14]). By the same principle according to which stars scintillate through irregularities in the earth's upper atmosphere, while planets, of finite angular diameter, do not, so very small radio sources will be observed to scintillate because of fluctuations in the electron density in the interplanetary medium.

Various kinds of estimates of time scales for radio emission can be obtained. The shortest time scale is that obtained by considering that the radio emission lying in extended, often double, volumes of space outside the parent galaxy has been caused by high-energy particles emitted in some explosive event in the parent galaxy. By assuming that the maximum extent of the radio source measures its true spatial extent, i.e., that there are no corrections due to projection on to the plane of the sky, and by assuming that the emission is due to high-energy particles which have been ejected at the speed of light, one finds from the dimensions of a few hundred kpc mentioned above, that we get time scales of about 3×10^5 years. We shall return later to further discussion of time scales.

III. **Identification of radio sources.** Sandage [21] has estimated that some 80% of the sources in the revised 3C catalogue published by the Cambridge radio astronomy group (Edge et al. [13]; Bennett [4]) can, and soon will be, identified with optical objects. The accuracy of radio positions is very rapidly being improved, by several independent groups of workers, so that the only sources which will remain unidentified will be those whose optical counterpart is too faint to be seen. As discussed in §I, it was originally thought that this fraction might be far larger than 20%. However, the great bulk of radio sources has proved not to be such strong emitters in the radio region as was first thought might be the case. That is to say, the frequency of radio sources per intensity interval increases very considerably from the brightest sources like Cygnus A to fainter sources.

In considering the types of optical objects that are identified with radio sources, we should first mention that there is a whole class of radio sources which lie within our own Galaxy. These mostly comprise supernova remnants; there is also a source at the center of our own Galaxy. It is the extragalactic radio sources that we are considering in this lecture. Of those that have been indentified, some 65% have proved to be elliptical, SO, and D-type galaxies (Matthews, Morgan and Schmidt [18]). D-type galaxies are very large elliptical galaxies with a greatly extended light distribution, so that they have a much larger diameter than ordinary elliptical galaxies. In this category, 50% of these identified galaxies are found in clusters, and they are always the brightest cluster member. Some 25% of the radio sources have proved to be quasi-stellar objects or N-type galaxies, none of which have been found to lie in clusters. N-type galaxies are galaxies with very bright nuclei, highly condensed, with only faint wispy outer parts. Five percent of the identifications have proved to be with double elliptical galaxies, called "dumbbells" by Matthews et al. [18], and the remaining 5% have been identified with spiral and irregular galaxies.

Our own Galaxy, of course, is a radio emitter, but only at a low level of power output—10^{38} erg/sec. We may compare this with the normal light emission of a bright galaxy of about 10^{44} erg/sec, and the radio output of the strongest sources like Cygnus A, of 10^{45} erg/sec. The existing radio source catalogues and surveys have been made down to some definite lower limit of radio intensity, set by the instrumental limitations. In the very near future, new catalogues will appear in which the lower limit has been extended greatly downwards. We may expect, then, that many normal galaxies like our own may be detected as radio emitters. Maybe every galaxy is a radio emitter at some level. Even if the emission were only contributed by an aggregate of supernova remnants, we should in fact expect this to be the case.

IV. **Mechanism of radio emission.** Alfvén and Herlofson [1] suggested that nonthermal radio emission might be produced by the synchrotron mechanism—high-energy electrons (and positrons) spiralling in a magnetic field. Shklovsky [24], noting that the synchrotron emission is highly polarized, suggested that, since the Crab Nebula is a radio source, the continuous optical radiation coming from it might also be produced by the synchrotron mechanism, and if so, this optical radiation would be highly polarized. Shortly after this, Dombrovsky and Baade discovered that in fact this radiation was polarized. Baade also showed that a curious jet of light coming from the radio galaxy M 87(Virgo A) was also highly polarized. These measurements confirmed in a striking way the theory that synchrotron emission is the source of the radio power output.

Accepting this mechanism as the correct one, we see right away that we have an enormous energetic problem to deal with. If the emitting particles have an energy spectrum:

$$N(E)dE = \text{constant} \times E^{-x}dE$$

up to a cut-off at energy E_{max}, the synchrotron spectrum is

$$S(\nu) = \text{constant} \times H_{\perp}^{(x+1)/2} \nu^{-(x-1)/2} f\left(\frac{\nu}{\nu_c}\right)$$

where ν_c depends on E_{max}; this gives a power output which is proportional to $\nu^{-\alpha}$ where $\alpha = (x - 1)/2$. Given the total power emitted (for which the distance of the radio source must be known), we can derive a total energy in particles as a function of the magnetic field $H_{\perp}$. Then, knowing the volume of the emission, we can also get the magnetic energy $\mathcal{M}$ as a function of $H_{\perp}$. It turns out that the energy in the particles and the energy in the magnetic field are minimized when they are equal. Total energies have been worked out (Burbidge [12]) for the case that there are protons dominating the high-energy flux of particles (it is only the electrons and positrons that will do the radiating, but it seems hard to produce a flux of these without also producing a flux of high-energy protons); the energies have also been calculated for the case when no protons are present. In either case, we arrive at very high energies that are necessary to explain the radio emission; the energy requirements are of course greater if protons are present. In fact, for the strongest radio sources, energies of up to $10^{60\text{-}62}$ ergs may be necessary.

In M 87, as just stated, optical synchrotron emission was detected; it has also recently been found in M 82 by Sandage and Miller [22]. The presence of optical synchrotron emission demands a renewed supply of high-energy radiating particles, or renewed acceleration of such particles, because the lifetimes of particles radiating in this wavelength region are very much shorter than the lifetimes of particles giving radio-frequency emission. We have the following expression for the half-life $\tau_{1/2}$ of a radiating electron:

$$\tau_{1/2} = \frac{8\cdot 4 \times 10^{-3}}{H^2 E} \qquad \text{years,}$$

where H is in gauss and E is in Gev. Now to produce radiation at optical frequencies, E has to be $\sim 2.5 \times 10^{12}$ ev for $H = 10^{-5}$. Thus lifetimes as short as about 30,000 years are involved.

V. **Detailed structure of radiogalaxies.** Let us now turn to a consideration of the detailed structure of radiogalaxies, in an attempt to determine just what is happening inside the galaxy to give rise to the radio emission. Now one would like to study the strongest of the radiogalaxies in order to do this, but objects like Cygnus A are rare in space and Cygnus A itself, the nearest of them, is so far away that little detail can be discerned. For detailed study, then, we have to choose galaxies which are not such strong radio emitters, but which are spatially more common so that we can choose nearby examples. I have selected a number of these objects, for which detailed studies are available, and we will consider them below. First, let us briefly consider the general appearance of the optical spectra of radio galaxies.

Characteristically, radiogalaxies show strong emission lines of hydrogen and forbidden lines of common light elements in several stages of ionization (somewhat like gaseous nebulae in our own Galaxy excited by a hot central star). However, in some cases the emission lines are not strong, and even in some radiogalaxies they are not present at all. When emission lines can be seen, measures of their intensities and Doppler displacements can be used to study both the conditions of excitation of the hot gas in the radiogalaxy, and the velocity field. When no emission lines can be discerned, we must presume that there is not much hot gas present in the radiogalaxy, and all the light that we see must be coming from stars.

Finally, before going on to discuss details of the selected radiogalaxies, a few words should be said about the source of the energy which somehow makes itself apparent in high-energy particles and magnetic fields and consequently in radio emission. The original suggestion, by Baade and Minkowski, was that the source of this energy was kinetic energy liberated in a collision between two galaxies. It was later found that this explanation would not work; firstly, many apparently single galaxies were in the category of strong radio sources, and also no satisfactory mechanism was found for transferring such collisional energy into the modes of high-energy particles and magnetic field. Later work, as we shall see, made it clear that the source of the energy appears to arise in the center of the galaxy. In fact, in the case of M 82, discussed below, the source of the energy was clearly in some violent explosive event right in the central region. Various explanations have been suggested, including the following: (i) enhanced supernova activity in a dense star region; (ii) a sudden collapse of material in the center leading to enormous numbers of collisions between stars; (iii) the formation of a massive object some $10^5 - 10^8$ solar masses, whose subsequent evolution might release both nuclear energy and rest mass energy.

M 82

This galaxy was the first object to give unequivocal evidence that a violent event had occurred in the nucleus of a galaxy (Lynds and Sandage [16]; Burbidge, Burbidge and Rubin [11]). This galaxy had long been known to be rotating, in the way that normal galaxies do, and to be oriented so that the line of sight makes only a small angle with the equatorial plane of the galaxy. Emission lines are quite strong in this galaxy, showing that it contains a large amount of excited gas. Lynds and Sandage obtained the spectrum along the minor axis or axis of rotation of the galaxy, and found that, where the component in the line of sight or rotational velocity should be zero, there was still a measured Doppler shift in the line, which was found to be proportional to distance from the center of the galaxy. The interpretation of this was that the gas along the minor axis had been exploded out of the galaxy and was expanding at a rapid rate, the fastest-moving gas having reached the furthest distance from the center. Consideration of the inclination of the galaxy, i.e., a discussion of which was the near side of the galaxy, showed that only an explosion and not a collapse of gas could explain the observations. The total energy emitted in the light of the hydrogen $H\alpha$ line was some 2×10^{40} ergs per second. The volume of the filaments which were emitting in $H\alpha$ light was estimated to be some $7 \times 10^{62}\,\mathrm{cm}^3$. From these two figures, a volume emissivity was derived from which the electron density could be determined. Lynds and Sandage found $N_e = 10/\mathrm{cm}^3$. The mass could also be derived, and was equal to $6 \times 10^6\,\mathrm{M}_\odot$; this, with the measured velocities, gave a kinetic energy of 2×10^{55} ergs. From the radio emission, by the usual calculations, assuming the synchrotron mechanism to be operating, that the magnetic and particle energies are equal, and that protons are present and dominate the flux of high-energy particles, a total energy of 10^{56} ergs was derived.

Later, Sandage and Miller [22] showed that the optical radiation in the region where the filaments exist was indeed highly polarized, confirming that this was optical synchrotron emission. A study of the excitation of the filaments that are emitting in $H\alpha$ light showed that the optical synchrotron emission, if extrapolated to the ultraviolet spectrum region, could provide sufficient radiation to ionize hydrogen and produce the recombination spectrum seen.

Lynds and Sandage derived a time scale for the explosion of 10^6 years; the later study by Burbidge, Burbidge and Rubin increased this time slightly to a few million years.

NGC 1275

Minkowski [19] found that the spectrum of this galaxy—which is the brightest galaxy in the Perseus cluster—showed a very interesting feature, namely, two sets of emission lines, with a velocity difference of 3000 km/sec between them which lent weight to the hypothesis current in 1957 that the object was the result of a collision between two galaxies. The nucleus of NGC 1275 is small and bright, with highly excited emission lines that are very broad and indicate a big velocity dispersion in the gas which is producing them. Galaxies with nuclei of this sort were discussed by Seyfert [23], and are now always designated as Seyfert galaxies.

Recently it has been shown that the velocity field in NGC 1275 indicates that a violent event has occurred in the center, as in M 82, rather than that a collision has taken place (Burbidge and Burbidge [9]). The redshift of the galaxy as a whole, as given by the velocity of its center, is 5300 km/sec. With a Hubble expansion constant of 75 km/sec per Mpc for the distance scale of the Universe, this gives a distance of 72 Mpc for NGC 1275, at which $1''$ corresponds to 350 pc. Thus this galaxy, being further away than M 82, cannot be studied in such detail as M 82. However, because it is a stronger radio source than M 82, it is interesting to find out as much as possible. Direct photographs taken mainly in the light of the $H\alpha$ emission line of hydrogen have shown an appearance suggesting narrow jets or bursts of gas emerging from the center of the galaxy in various directions. The detailed study of the velocity field, from measurements of the Doppler shift in various position angles all around the galaxy, also suggests this interpretation. Now the displaced velocity component, discovered by Minkowski, has a magnitude such that the recession velocity of this gas is larger than that of the galaxy as a whole. Thus the gas giving rise to this component must lie behind the galaxy, if this gas was in fact ejected from the center of the galaxy. It is interesting that the velocity dispersion in this ejected gas is quite small. This suggests a steady or intermittent ejection, rather constant in direction and velocity, over a time scale which can be estimated from the extent of the gas giving rise to the displaced velocity component; this time scale is 5×10^6 years.

The main difference between NGC 1275 and M 82 is that NGC 1275 is larger, almost certainly more massive, and has a larger stellar component. The disturbed velocities measured in the excited gas in NGC 1275 are also larger, and a simple pattern of ejection, with velocity proportional to distance from the center, which was found in M 82, cannot be deduced in this case. An estimate for the kinetic energy, made approximately as was done for M 82, is 10^{58} ergs. From the observed radio power output, the energy in the particles and magnetic field has to be about 10^{59} ergs. Thus the magnitude of the violent event that must have given rise to this radio galaxy is altogether greater than was the case in M 82.

NGC 1068

This galaxy is a fairly weak radio emitter, but it is comparatively near to us, and so can again be studied in some detail. It also is a Seyfert galaxy, with a small bright nucleus whose spectrum shows strong emission lines of high excitation which are very broad. The mass of the galaxy is only a few times 10^{10} solar masses; this leads to to a velocity of excape from the central region of the galaxy that is only about 400 km/sec. Now the breadth of the emission lines coming from the nucleus gives a range of velocity considerably larger than this, running up to about 2250 km/sec. Thus this gas cannot be contained by the galaxy, and must be escaping from it with a time scale which will be given simply by the dimension of the nucleus divided by the velocity. Measures of the diameter of the bright central nucleus suggest a size of 50 pc, and this, with a velocity of 2250 km/sec, gives a time scale of 2×10^4 years. In fact, Walker [26] has recently shown that the broad emission lines have structure in them, suggesting that they come from fast-moving clouds whirling about in the center of the galaxy.

Osterbrock and ·Parker [20] have studied the intensities of the emission lines to deduce the physical conditions in the gas right in this central region. They have determined the electron density, N_e, from the $H\alpha$ flux as was done in the case of M 82; then, since the number of protons must equal the number of electrons, and hydrogen is the most abundant element, this leads to a total mass of 2×10^6 solar masses. But with this mass, and the dimension of 50 pc, the electron density comes out much lower than that deduced from study of the intensities of the emission lines. This again shows that the gas must be located in discrete clouds, and does not fill the whole volume of the nuclear region. The real mass of the ionized hydrogen, derived by Osterbrock and Parker, is 3×10^4 solar masses.

These authors also studied the source of the ionization of the hot gas in the center. They showed that it could not be due to electron collisions, nor to ultraviolet radiation from a single central source as is the case in gaseous nebulae around hot stars in our own Galaxy, nor could it be due to ultraviolet synchrotron emission as is the case in M 82. They concluded that the best mechanism for ionization was through collisions with fast protons. By fast protons, I do not mean protons moving at relativistic speeds which might be co-existent with the electrons giving the synchrotron radio emission. Osterbrock and Parker showed that the most likely value for the energies of the protons was 20 kev, and this corresponds exactly to protons moving with velocities around 2000 km/sec. In other words, the source of ionization can be the collisions of the clouds, in which kinetic energy is converted to ionization energy. This does not, of course, tell us what is the source of the kinetic energy; it merely puts this problem one stage further back.

The radio power output in NGC 1068 is 7×10^{39} erg/sec; the kinetic energy in the gas is 1.2×10^{54} ergs, and the energy in the radio source contained in particles plus field (with protons present) is 4×10^{55} ergs.

VI. **Concluding remarks** There are other radio galaxies which have been studied in less detail than those considered above; NGC 5128 (Burbidge and Burbidge [7], Bolton and Clarke [5], Burbidge and Burbidge [8]), NGC 1316 (Arp [2]), NGC 6166 (Burbidge [6]), and NGC 4782-3 (Burbidge, Burbidge, and Crampin [10]) are other examples, but we do not have time to consider these in detail. In any case, in none of these galaxies are the emission lines as strong as in the ones that we have discussed in detail. Emission lines are visible in the first three of the objects listed above, but none at all can be detected in NGC 4782-3, and, as we have seen, unless emission lines are present and extensive in a radiogalaxy, we cannot use them to study kinematics and excitation conditions in the hot gas.

In conclusion, we may sum up the main observational facts which a successful theory of radiogalaxies has to explain. First, large sources of energy are required, and, second, this energy has to be in the modes of relativistic charged particles and magnetic fields. These are the main points. Then we note that a characteristic feature of the strong radiogalaxies is that they are mostly massive elliptical galaxies, the bulk of whose matter is condensed into old stars; from this point of view any hypotheses suggesting that young galaxies in the process of formation might be involved will not work. Next, we have to explain the rather common feature of a double radio structure; this suggests a preferred axis of ejection of charged particles, and in the cases of the nearby source NGC 5128 and a distant radiogalaxy 3C 33, observed by Schmidt, the long axis of the radio distribution appears to lie close to the axis of rotation of the galaxy.

Finally, there is the problem of time scales and the necessity for events giving rise to the radio emission to be able to occur more than once in a galaxy. This is shown in the structure of NGC 5128, which has a small double radio source within the galaxy as well as the enormously extended radio distribution outside it. This is also the case in any radiogalaxy which has optical synchrotron emission as well as an extended outer radio source, as does M 87 (Virgo A), because of the short lifetimes of the electrons radiating in the optical wavelength region.

Some people claim that they are not interested in devising a theory to explain too great a collection of observed facts; as an observer myself—a collector of facts—I would rather suggest that such a task is more challenging and therefore more exciting!

References

1. H. Alfvén and N. Herlofson, Phys. Rev. 78(1950), 616.
2. H. Arp, Astrophys. J. 139 (1964), 1378.
3. W. Baade, and R. Minkowski, Astrophys. J. 119 (1954), 206.
4. A. S. Bennett, Mem. Roy. Astr. Soc. 68(1962), 163.
5. J. G. Bolton, and B. G. Clark, Publ. Astr. Soc. Pacific 72 (1960), 29.
6. E. M. Burbidge, Astrophys. J. 136 (1962), 1134.
7. E. M. Burbidge and G. R. Burbidge, Astrophys. J. 129 (1959), 271.

8. _____, Nature **194**(1962), 367.

9. _____, Astrophys. J. **142** (1965), 000.

10. E. M. Burbidge, G. R. Burbidge and D. J. Crampin, Astrophys. J. **140** (1964), 942.

11. E. M. Burbidge, G. R. Burbidge and V. C. Rubin, Astrophys. J. **140** (1964), 942.

12. G. R. Burbidge, Paris Symposium on Radio Astronomy, IAU Symposium No. 9 (Ed. R. N. Bracewell, Stanford Univ. Press) 1959; p. 541.

13. D. O. Edge, J. R. Shakeshaft, W. B. McAdam, J. E. Baldwin and S. Archer, Mem. Roy. Astr. Soc. **68**(1959), 37.

14. A. Hewish, P. F. Scott and D. Wills, Nature **203**(1964), 1214.

15. R. C. Jennison and M. K. Das Gupta, Nature **172**(1953), 996.

16. C. R. Lynds and A. R. Sandage, Astrophys. J. **137** (1963), 1005.

17. P. Maltby and A. T. Moffett, Astrophys. J. Suppl. **7** (1962), 141.

18. T. A. Matthews, W. W. Morgan and M. Schmidt, 1964, Proc. Dallas Symposium on Quasi-Stellar Sources and Gravitational Collapse, Univ. Chicago Press, Chicago, Ill., 1964; p. 105.

19. R. Minkowski, 1957, Radio Astronomy, IAU Symposium No. 4, Ed. H. C. van de Hulst, Cambridge Univ. Press, New York 1957; p. 107.

20. D. E. Osterbrock and R. A. R. Parker, Astrophys. J. **141** (1965), 892.

21. A. R. Sandage, Lectures given at Varenna Summer School on High-Energy Astrophysics, 1965.

22. A. R. Sandage and W. C. Miller, Science **144**(1964), 405.

23. C. K. Seyfert, Astrophys. J **97**(1943), 28.

24. I. S. Shklovsky, Dokl. Akad. Nauk. SSSR **90**(1953), 983.

25. F. G. Smith, Nature **168**(1951), 555.

26. M. F. Walker, Ann. Rept. Lick Obs., Astronom. J. **68** (1963), 643.

UNIVERSITY OF CALIFORNIA, SAN DIEGO

LA JOLLA, CALIFORNIA

John Linsley

Cosmic Radiation

First lecture. My own work has been for some time with the high energy portion of the cosmic ray spectrum, where the energy per particle is 10^{15} eV or more. Before that I did some work on primary radiation at energies near the geomagnetic cutoff. The sun was quiet then, and by hindsight we knew we were dealing with galactic radiation, pretty much undisturbed by anything in the solar system except the earth's field. That was at Minnesota. So now I shall consider natural corpuscular radiation in space with energy, or energy per nucleon, greater than 10^{10} eV. I will leave out modulation effects. As for solar cosmic rays, I will be concerned only with properties that set them off as a distinct phenomenon. Thus I will keep out of the territory in which cosmic ray physics overlaps physics of the solar system.

Next I consider the relation of cosmic radiation, as a subject, to astronomy. One can put it this way: astronomers are after information concerning distant objects and activities, information that is carried here by just two vehicles, electromagnetic radiation and corpuscular radiation. If I do put it that way I must then be very apologetic, because the corpuscular radiation carries very little information. It has a certain *composition*, each component has an *energy spectrum* and a *directional distribution*, and there is the independent variable *time*. But the energy spectra seem to have very little structure, the radiation seems to be isotropic in direction, and there seem to be no time variations. No matter, and no need for apology, we are after meaning. Information is means to an end. The statement "the sun is a star" contains very little information in a formal sense but is very rich in meaning. Cosmic radiation says something significant in few words.

The significance has to do with energy and temperature. The temperatures are extraordinarily high. A typical particle energy, 10 GeV, corresponds to 10^{14} degrees, much greater than the temperature of nucleons in a nucleus, and the range extends much higher. It is surprising that any detectable amount of matter gets into such a state.

Perhaps after all the amount that does is relatively very small. That would be true if ordinary quiet stars were the source of cosmic radiation. We live very near the sun, which is such a star, ordinary. The energy density of cosmic radiation is less than that of sunlight by a factor 10^{13}, so the heating could be accomplished by a process having very small intrinsic probability. To help with the explanation there is the effect of magnetic fields, which retard outflow of charged particles. Thus energy density can build up near a source, unlike the situation with light. On the other hand, if the sun is not the main source of our local corpuscular radiation, then there is much more to be explained, and less with which to explain it. More to be explained, because the local energy density must then be considered typical for the whole galaxy. The comparison must be with starlight, not sunlight, which means there is no factor of 10^{13}. Even that comparison, between the energy density of cosmic radiation and that of starlight, would fail to tell the story because most of the starlight comes from stars more or less similar to the sun, stars that would have to be ruled out as sources if the sun were ruled out.

What, then, is the evidence against solar origin? Isotropy is not. Particles might wander about within a certain "sphere of influence" of the sun long enough so that their directions would be random by the time they were observed. Emission by the sun might take place continually at a very slow rate or there might be brief intense bursts during which isotropy is disturbed, after which it returns.

It does not help very much, either, to point out that *some* individual particles could not have been accelerated by the sun because their individual energy is too great. Particles that could not have come from the sun for *that* reason make practically no contribution to the energy density. They could be attributed to processes that are trivial as consumers of energy, however interesting they might be for other reasons.

Convincing evidence was obtained by discovering that the sun does indeed emit corpuscular radiation, and that "solar cosmic rays"—identified by association with definite flares on the sun—are distinctly different than "ordinary cosmic rays." For one thing the solar particles have a much steeper energy spectrum. In effect it extends only to about $30\,\mathrm{GeV}$. Even more conclusive is the difference in composition. For example, the ratio of helium nuclei to those in the charge group $11 \leq Z \leq 18$ is $(800 \pm 90):1$ in solar cosmic rays as compared to $(48 \pm 7):1$ in the radiation *not* associated with flares, or different by a factor 16 ± 3.[1] The composition of solar cosmic rays is like that of the solar surface or the surface of most stars. "Ordinary" (i.e., galactic) cosmic rays are peculiarly rich in nuclei of the heavier elements.

Current opinion favors the view that galactic cosmic rays originate in supernova explosions. The peculiar composition I just spoke of supports that explanation, and the synchrotron radiation emitted by supernova remnants indicates that high energy particles are actually present, that acceleration is still going on centuries after the explosive phase. The particles that radiate are electrons of energy 10^9 to $10^{12}\,\mathrm{eV}$, but presumably there are also energetic protons and other nuclei. Finally, there is enough energy release per explosion, and explosions are frequent enough, to maintain the observed energy density of cosmic rays if one assumes an efficiency of a few percent or more. There is the explosive phase; there is continued acceleration in the turbulent magnetic fields of the remnant; and there may be additional gain in energy through collisions with interstellar magnetic clouds. There are differences of opinion as to the relative gains made in each of the three stages. According to this theory injection takes place at infrequent times

---

[1] S. Biswas and C. E. Fichtel, NASA Publication X-11-65-219, May 1965 (submitted for publication to Space Science Reviews).

and at discrete locations; nevertheless we find isotropy and no large time variations, even over rather long intervals. That means good mixing in a reservoir of large capacity.

It should be noted that isotropy and time-independence inform as to *propagation*, not *acceleration*. They tell about the strength and character of interstellar magnetic fields, that they are chaotic, not regular, so there is diffusion, not orderly streaming. We are more likely to learn something about the *acceleration* process from studies of *composition as a function of energy*, especially near the low and high ends of the spectrum. One might hope for hints as to how conditions unfold during the explosion itself. At the start there might be composition A with spectrum B; at the end it might be composition X with spectrum Y. According to present evidence the spectra of different nuclei are remarkably similar, but the range covered by good measurements is small. Most of this work has been done at comparatively low energies where there is modulation by fields associated with the sun. It will be hard to draw conclusions about the character at the source until this modulation is better understood.

Second lecture. Earlier I touched on solar cosmic rays and galactic cosmic rays. Now I hope to discuss extragalactic cosmic rays, but first I should like to amplify some of my earlier remarks.

About the measurements, most of the results on low energy nuclei have come from use of photographic emulsion. Typically, a stack of pure emulsion in layers 600 microns thick is flown with a balloon for 8 or 10 hours. The thickness of residual atmosphere above the balloon may be as small as $3.5 \, \text{g}/\text{cm}^2$, or sometimes even less. For precision work on composition the volume of the stack may be 3 liters. For work on high energy interactions the volume has been as great as 80 liters. For study of solar cosmic rays one may use a very small stack (because the intensity is great) and a rocket. Measureable quantities are

1. rate of energy loss by ionization, $- dE/dx$, given by grain density, gap density, or density of delta-rays.

2. mean square angle of multiple Coulomb scattering, for a given "cell thickness" or increment of thickness along the track, denoted by $\langle \theta^2 \rangle$.

3. range, denoted by R.

4. magnetic rigidity, from the geomagnetic cutoff and latitude

of the exposure. Magnetic rigidity is momentum per unit charge, p/z. The following equations relate those quantities to z, the charge in units of e; to A, the rest mass in units of the proton rest mass; and to γ, the Lorentz factor $(1 - v^2/c^2)^{-1/2}$:

(1) $$-dE/dx = z^2 f_1(\gamma),$$

(2) $$\langle \theta^2 \rangle = (z/A)^2 f_2(\gamma),$$

(3) $$R = (A/z^2) f_3(\gamma),$$

(4) $$P/z = (A/z) f_4(\gamma).$$

The quantities f_1 through f_4 are known functions. For low energies ($\gamma \overset{\sim}{<} 2$) they differ in form enough so that if there are 3 measurements one can determine all three quantities; A, z and γ. We expect z to be an integer and A to be either an integer or $1/1840$. I leave it an exercise to design an experiment to tell whether the primaries arrive stripped of electrons. Note that the earth's field acts *before* particles traverse an appreciable thickness of atmosphere.

To show how far the emulsion technique has been pushed, there is a result by Hildebrand and others at NRL giving the ratio of He^3 to He^4 as

$$\left(0.06 \begin{array}{c} + 0.03 \\ - 0.02 \end{array} \right).$$

Considering that the mass ratio is only $4:3$, and that one component was much more abundant than the other, it was quite a feat.

As for electrons and gamma rays, in both cases there is competition between those that arrive from outside the vicinity of the earth and those produced locally by the far more abundant nuclei. The convincing evidence for primary electrons has been obtained with cloud chambers and spark chambers. The electrons, which had energies of order $1\,GeV$, were identified by the distinctive cascades they produce. The abundance was about 1% compared to protons of the same energy. To separate positive and negative electrons a magnet was used. In mentioning earlier the result, that both signs were found with negatives more abundant, I said that the negative excess means that some electrons are "primary." I should warn you that the terms "primary" and "secondary" are used ambiguously now. They are still used in the original sense to distinguish between particles arriving from outside the vicinity of the earth and those that come from interactions in the atmosphere or other local material. The terms are also used to distinguish between electrons that were *accelerated* to their observed state by the same large-scale process that accelerates nuclei, and electrons produced by high energy nuclei through a collision followed by π-μ-e decay, a collision that takes place far from our solar system, during propagation or perhaps in an acceleration region.

Cosmic gamma rays are expected from decay of neutral pions produced in collisions by nuclei, and from collisions of electrons with photons by inverse Compton effect. They would be "secondary" in the new sense. A line is expected at 0.5 MeV from annihilation of positrons, and there is a possibility, in the region of a few hundred MeV, that annihilation of antimatter might contribute. Vigorous efforts have been made to detect gamma rays, but it is not certain whether any of them has succeeded yet. It was expected that "primaries," in the old sense, would be anisotropic, but whatever is being measured comes out to be isotropic. The experimental problem is to discriminate strongly enough against secondaries (old sense) produced in the measuring equipment of the earth's atmosphere. One of the best experiments was done using a satellite, Explorer XI. The observed intensity was about 0.1% of the total cosmic ray intensity, so we have that as an upper limit.

Next I would like to give some flavor of argumentation as to energetics. The central *fact* is local energy density. What we observe directly is intensity in a particular direction, the vertical. This is done at the top of the atmosphere, at a high latitude where not much is cut out by the earth's field, and at a favorable time in the solar cycle. The result is about 0.3 nucleons/(cm^2 steradian sec). Protons are counted as 1, helium nuclei as 4, and so on. The contribution of heavies is less than half, but appreciable. Next, from the energy spectrum we derive the mean energy per nucleon, which comes out about 5 GeV. Multiply 0.3 by 4π sterad; then multiply by 5×10^9 eV; then divide by 3×10^{10} cm/sec, the velocity. The result is about 0.6 eV/cm^3, or 10^{-12} erg/cm^3.

Illustration. To get the *minimum* power requirement, assuming uniform cosmic ray density in this galaxy, multiply 10^{-12} by the volume of the galaxy, 10^{68} cm^3, and divide by the maximum permissible time allowed for an average particle to sustain its contribution. One needs an educated guess as to the average density of material encountered in travel. As an example, Ginzburg has used 2×10^{-26} g/cm^3, or one proton per 100 cm^3. (This is 100 times less than the density in the disk.) The *maximum time* is that corresponding to traversal of 3 g/cm^2, the mean free path for collisions by iron nuclei in hydrogen. Otherwise there would be no Fe left. On the other hand, 3 g/cm^2 is itself acceptable. Starting with a goodly proportion of heavy nuclei, or even pure Fe, we would get enough of the light elements Li, Be, B to agree with observation. Using those figures the maximum time comes out 5×10^{15} sec, or 2×10^8 years. The minimum power would be, therefore, 2×10^{40} erg/sec.

The energy requirement is met by the supernova theory as follows, quoting Ginzburg rate of explosions 1 in 30 to 100 years, mean energy given to cosmic radiation per explosion 1 to 3×10^{49} erg. By simple division one gets something of order 10^{40} erg/sec, which Ginzburg feels can be pushed upward as much as one order. He supports the assumption as to energy given to cosmic radiation by the claim that total energy release is generally more than 10^{50} erg and sometimes is two orders greater.

Burbidge says, in opposition, that "The total input of energy per supernova is quite uncertain," and points to how little is known about supernovae from actual observations. "From a theoretical standpoint the total energy which could be released·· ·is·· ·perhaps $10^{51} - 10^{52}$ erg per solar mass, but it is not clear whether the energy that is actually released lies in this range or is several orders less than this, nor do we know what fraction goes into high-energy particles." Each disputant raises other issues as well; I present this merely to illustrate usage.

Now I turn to a different notable characteristic, the extreme magnitude of the *energy per particle*, or the extreme *temperature*. I can express the "accepted view" up to a few years ago by quoting B. Peters (speaking, actually, in retrospect), "—the search was guided by the idea that we have to confine ourselves to the galaxy because we cannot hope to find sources powerful enough to fill the whole universe, and in any case it would be unthinkable that the energy in the cosmic radiation would be so high as to be comparable with all other forms of energy in the universe except its rest-mass." That was the context. The key word is of course "unthinkable." Ginzburg wrote in 1960, "even when $E = 5 \times 10^{17}$ eV/nucleon the condition· ·which is necessary for the acceleration, can be fulfilled· ·." That was in support of the supernova theory as adequate to explain everything that had been observed. The highest energy *per nucleus* he had read about when he wrote was a few times 10^{19} eV, so by assuming the particle was an iron nucleus he could just make it. At the same conference we of MIT reported a new record, 6×10^{19} eV, and a short time later the present record, 10^{20} eV/nucleus, or at least 2×10^{18} eV/nucleon. A parallel quotation from 1963 reads, "The appearance of particles with energies $\geq 10^{17}$ eV/nucleon as a result of the flares of supernovae is hardly possible." Considering his English, which is delightful but not always clear, this could mean "it is possible but difficult," or "it is not possible at all." Peters, another die-hard, also concedes now that it is impossible to explain the highest energy with supernovae. Both men are attracted by a new possibility, that violent activity on a much larger scale than in supernovae may have occured, or be taking place now, in the nucleus of this very galaxy. I think it is accepted by all that even if they were produced here the highest energy particles would not stay, so discussion has begun concerning the energy density of cosmic radiation in the space between galaxies, the so-called "metagalactic space."

A somewhat extreme view is that of Burbidge. He notes that within a distance of 10^{27} cm (volume 10^{81} cm^3) there are 10^3 strong radio sources. In the "age of the universe" there have been 10^{10} years for the intensity to build up. He gives evidence that the lifetime of a radio source may be only 10^6 years, so there may have been contributions from 10^7 of them. Using two assumptions, that the average field in a radiosource is $\overset{\sim}{<} 5 \times 10^{-6}$ Gauss, and that the number of protons is 100 times the number of electrons, plus observational data, he claims the energy input to particles may be as great as 10^{62} erg per event. This is pushing hard, the assumptions have the effect of maximizing the energy given to particles. In that way, however, he can get an average density in the metagalaxy of 10^{-12} erg/cm^3, the same as that near the earth. He admits that for the same mechanism 10^{-14} erg/cm^3 is a more probable figure, for cosmic ray energy density in the metagalaxy. He points out that if the ratio (energy belonging to protons/energy belonging to electrons) = 100 is assumed in the metagalaxy as well as at production, then with total energy density 10^{-14} erg/cm^3 there would be enough electrons to give the observed gamma rays (by inverse Compton effect) assuming that what has been observed is not artifact.

Opposing views are expressed by Ginzburg. He points out that leakage from *normal* galaxies would by now have built up an energy density in the metagalaxy of 10^{-16} to 10^{-17} erg/cm^3. If it were much greater, and if 1% or more belonged to electrons, then there would be detectable synchrotron radiation, contrary to observation, assuming of course some "reasonable" value for the magnetic field strength. And he maintains that the energy density of cosmic radiation ought not to be greater than that of other forms; e.g., that of turbulent gas motion, which he estimates to be 10^{-14} to 10^{-15} erg/cm^3.

MASSACHUSETTS INSTITUTE OF TECHNOLOGY
CAMBRIDGE, MASSACHUSETTS

Index